Advanced Computational Methods and Experiments in Heat Transfer X

WIT*PRESS*

WIT Press publishes leading books in Science and Technology.
Visit our website for the current list of titles.
www.witpress.com

WIT*eLibrary*

Home of the Transactions of the Wessex Institute.
Papers presented at Heat Transfer 2008 are archived in the WIT elibrary in volume 61 of WIT Transactions on Engineering Sciences (ISSN 1743-3533).
The WIT electronic-library provides the international scientific community with immediate and permanent access to individual papers presented at WIT conferences.
http://library.witpress.com

TENTH INTERNATIONAL CONFERENCE ON ADVANCED COMPUTATIONAL METHODS AND EXPERIMENTAL MEASUREMENTS IN HEAT TRANSFER

HEAT TRANSFER X

CONFERENCE CHAIRMEN

B. Sundén
Lund University, Sweden

C.A. Brebbia
Wessex Institute of Technology, UK

INTERNATIONAL SCIENTIFIC ADVISORY COMMITTEE

Organised by
Wessex Institute of Technology, UK
Lund University, Sweden

Sponsored by
The Development in Heat Transfer Book Series

WIT Transactions

Transactions Editor

Editorial Board

Advanced Computational Methods and Experiments in

Heat Transfer X

Editors

B. Sundén
Lund University, Sweden

C. A. Brebbia
Wessex Institute of Technology, UK

WIT*PRESS* **Southampton, Boston**

B. Sundén
Lund University, Sweden

C.A. Brebbia
Wessex Institute of Technology, UK

Published by

WIT Press
Ashurst Lodge, Ashurst, Southampton, SO40 7AA, UK
Tel: 44 (0) 238 029 3223; Fax: 44 (0) 238 029 2853
E-Mail: witpress@witpress.com
http://www.witpress.com

For USA, Canada and Mexico

Computational Mechanics Inc
25 Bridge Street, Billerica, MA 01821, USA
Tel: 978 667 5841; Fax: 978 667 7582
E-Mail: infousa@witpress.com
http://www.witpress.com

British Library Cataloguing-in-Publication Data

A Catalogue record for this book is available
from the British Library

ISBN: 978-1-84564-122-1
ISSN: 1746-4471 (print)
ISSN: 1743-3533 (online)

The texts of the papers in this volume were set individually by the authors or under their supervision. Only minor corrections to the text may have been carried out by the publisher.

Printed and bound in Great Britain by Cambridge Printing

Preface

Research and developments of computational methods for solving and understanding heat transfer problems continue to be important because heat transfer topics are commonly of a complex nature and different mechanisms like heat conduction, convection, turbulence, thermal radiation and phase change may occur simultaneously. Typically, applications are found in heat exchangers, gas turbine cooling, turbulent combustion and fires, electronics cooling, melting and solidification etc. Heat transfer might be regarded as an established and mature scientific discipline, but it has played a major role in new application fields such as sustainable development and reduction of greenhouse gases as well as for micro- and nanoscale structures and bio-engineering. Non-linear phenomena other than momentum transfer may occur due to temperature-dependent thermophysical properties. In engineering design and development, reliable and accurate computational methods are requested to replace or complement expensive and time consuming experimental trial and error work. Tremendous advancements have been achieved during recent years due to improved numerical solutions of non-linear partial differential equations and computer developments to achieve efficient and rapid calculations. Nevertheless, to further progress in computational methods will require developments in theoretical and predictive procedures – both basic and innovative – and in applied research. Accurate experimental investigations are needed to validate the numerical calculations.

Many of the research topics were discussed during the Tenth International Conference on Advanced Computational Methods and Experimental Measurements in Heat Transfer held in Maribor, Slovenia in July 2008. The objective of this conference series is to provide a forum for presentation and discussion of advanced topics, new approaches and application of advanced computational methods and experimental measurements to heat and mass transfer problems. This book contains the edited versions of the papers presented at the Conference. All papers have been reproduced from material submitted by the authors but an attempt has been made to use a unified outline and presentation for each paper.

The editors would like to thank all the distinguished and well-known scientists who supported our efforts by serving on the International Scientific Advisory Committee, reviewing the submitted abstracts and papers. The excellent administrative work of the conference secretariat at WIT is greatly appreciated and the efficient co-operation and encouragement by the staff at WIT Press contributed significantly in producing this excellent conference book.

Bengt Sundén and Carlos Brebbia
Maribor, 2008

Contents

Section 1
Natural and forced convection

Numerical investigation of transient single phase forced convection of nanofluids in circular tubes

V. Bianco, O. Manca & S. Nardini
Dipartimento di Ingegneria Aerospaziale e Meccanica,
Seconda Università degli Studi di Napoli,
via Roma 81031Aversa (CE), Italy

Abstract

In this paper the development of the laminar forced convection flow of a water–Al_2O_3 nanofluid in a circular tube submitted to a constant and uniform heat flux at the wall is numerically investigated in a transient regime. A single and two-phase model (discrete particles model) is employed with either constant or temperature-dependent properties. The investigation is accomplished for a particle size equal to 100 nm. The convective heat transfer coefficient for nanofluids is greater than that of the base liquid. Heat transfer enhancement increases with the particle volume concentration, but it is accompanied by increasing wall shear stress values.
Keywords: nanofluid, forced convection, transient numerical analysis.

1 Introduction

Convective heat transfer enhancement is a continuous demand in many industrial heating or cooling equipment. An innovative technique for improving heat transfer by using ultra fine solid particles in the fluids has been used extensively during the last decade. Maxwell [1] showed the possibility of increasing thermal conductivity of a mixture by greater volume fraction of solid particles. These fluids, containing colloidal suspended nanoparticles, have been called nanofluids. Nowadays there is a fast growth of research activities in this heat transfer area, as recently reviewed for example in [2–4].

www.witpress.com, ISSN 1743-3533 (on-line)
doi:10.2495/HT080011

Different concepts and models have been proposed to explain the enhancement in heat transfer [5,6]. Experimental results were obtained on convective heat transfer for the laminar and turbulent flow of a nanofluid inside a tube in [5, 7]. Experimental results for the convective heat transfer of Al_2O_3 (27–56 nm)/water based nanofluids flowing through a copper tube in a laminar regime were reported in [8]. Numerical investigations on nanofluids have been carried out by two approaches. The first approach assumes that the continuum assumption is still valid for fluids with suspended nanosize particles. The other approach uses a two-phase model for better description of both the fluid and the solid phases. Another approach is to adopt the Boltzmann theory. The single phase model with physical and thermal properties all assumed to be constant with temperature was employed in [9–13].

The hydrodynamic and thermal characteristics of nanofluids flowing through a uniformly heated tube in both laminar and turbulent regimes with adjusted properties were investigated in [9]. The disadvantages of nanofluids with respect to heat transfer were discussed [10]. It was found that the inclusion of nanoparticles introduced drastic effects on the wall shear stress. A new correlation was proposed in [11] to describe the thermal performance of Al_2O_3-water nanofluids under turbulent regime. A numerical study of heat transfer for water–Al_2O_3 nanofluids in a radial cooling system was accomplished in [12]. They found that the addition of nanoparticles in the base fluids increased the heat transfer rates considerably. Laminar forced convection flow of nanofluids between two coaxial and parallel disks with central axial injection has been considered using temperature dependent nanofluid properties in [13]. The single phase flow model was solved numerically. A numerical study on the fully developed laminar mixed convection of a nanofluid consisting of water and Al_2O_3 in a horizontal curved tube was carried out in [14]. Three-dimensional elliptic governing equations were used and the single phase model was employed. The two phase approach seems a better model to describe the nanofluid flow. In fact, the slip velocity between the fluid and particles may not be zero [5] due to several factors such as gravity, friction between the fluid and solid particles and Brownian forces, the phenomena of Brownian diffusion, sedimentation and dispersion. Very recently, two phase mixture models were applied to study the turbulent forced convection flow of a nanofluid in a uniformly heated tube [15] and in the laminar mixed convection of a nanofluid in a horizontal tube [16]. A comparison between single phase and two phase approaches was accomplished in terms of temperature and velocity distributions and Nusselt number profiles in [17]. The comparison was carried out for steady state developing laminar forced convection flow in a circular tube heated at uniform heat flux.

In this paper the transient laminar forced convection flow of a nanofluid in a circular tube is numerically investigated. A two dimensional axial symmetric flow is considered and the circular tube is heated at uniform heat flux. The study is carried out for water with alumina particles with a spherical size of 100 nm diameter. The CFD commercial code Fluent is employed to solve the problem by means of the finite volume method. Single phase and two phase approaches are employed to evaluate the developing laminar forced convection flow. A comparison between results obtained by two different models is accomplished in terms of temperature and velocity distributions and the surface heat transfer coefficient.

2 Mathematical modelling

Figure 1 shows the geometrical configuration under consideration. It consists of the transient, forced laminar convection flow and heat transfer of a nanofluid flowing inside a straight tube of circular cross-section. The basic fluid is water and the particles are of Al_2O_3. The tube length is L=1.0 m and it has a diameter $D=1.0\times10^{-2}$ m. The fluid enters at uniform temperature and axial velocity profiles at the inlet section. The condition of the axially and circumferentially uniform wall heat flux is considered in this study. Also, the flow and the thermal field are assumed to be symmetrical with respect to the vertical plane passing through the tube main axis and the half tube is considered.

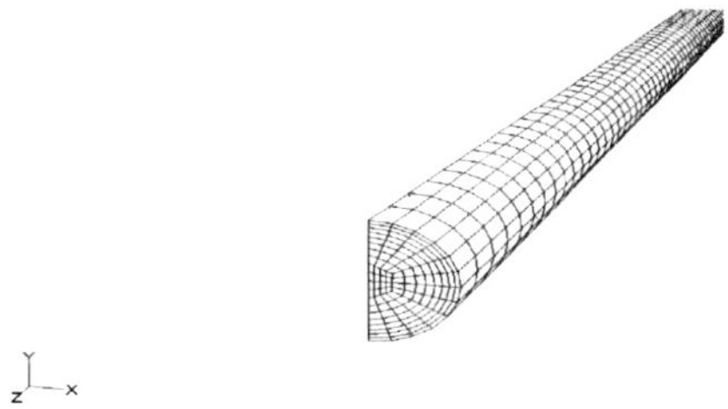

Figure 1: Geometrical configuration under study.

The single phase model, which has been used frequently for nanofluids, is also implemented to compare its predictions with the mixture model. The equations reported in table 1 represent the mathematical formulation of the single phase model.

Table 1: Governing equations.

Mass conservation:	Momentum conservation:	Energy conservation:
$\nabla\cdot(\rho\vec{V})=0$	$\rho\frac{D\vec{V}}{Dt}=-\nabla p+\mu\nabla^2\vec{V}$	$\rho c\frac{DT}{Dt}=\nabla\cdot(k\nabla T)$

The compression work and the viscous dissipation are assumed negligible in the energy equation.

The same equations are solved for the continuous phase of the two-phase model. Discrete phase is made of spherical particles following the model employed in the Fluent code [18]. The model is given by Ounis et al. [19].

At the tube inlet, profiles of uniform axial velocity V_o (it is calculated from Re, according to the fluid properties) and temperature T_o (293K) prevail. At the tube exit section, the fully developed conditions prevail, that is to say that all axial derivatives are zero. On the tube wall, the usual non-slip conditions are imposed; and one thermal boundary condition has also been considered in this study, namely the uniform wall heat flux. As noted earlier, both the flow and thermal fields are assumed symmetrical with respect to the vertical plane passing through the tube main axis.

The determination of nanofluid properties is, as previously mentioned, at the center of current nanofluid research. For the single phase model the following formulas were used to compute the thermal and physical properties of the nanofluids under consideration (the subscripts p, bf and nf refer to the particles, the base fluid and the nanofluid, respectively).

Density: In the absence of experimental data for nanofluid densities, constant-value temperature independent densities based on nanoparticle volume fraction are used:

$$\rho_{nf} = (1-\varphi)\rho_{bf} + \varphi\rho_{bf} \tag{7}$$

Specific heat: Similarly, in the absence of experimental data relative to nanofluids, it has been suggested that the effective specific heat be calculated using the following equation:

$$Cp_{nf} = (1-\varphi)Cp_{bf} + \varphi Cp_{p} \tag{8}$$

Dynamic viscosity: In this work in the first case, dynamic viscosity dependent only on φ is considered, then in the second case the variability with the temperature is also considered.

In the first case the dynamic viscosity of nanofluids has been obtained by performing a least-square curve fitting of some scarce experimental data available for the mixtures considered [20–22].

In the second case it is assumed that variable nanofluid properties (with temperature) will yield even better performance predictions than when considering constant properties. In this present work, we have used only experimental data published by [23] as the basis for the properties specified in the numerical model.

Thermal conductivity: As dynamic viscosity in the first case, thermal conductivity dependent only on φ is considered, then in the second case the variability with the temperature is also considered.

In the first case the model proposed by [24] has been used, assuming spherical particles. In the second case it is assumed that variable nanofluid properties (with temperature) will yield even better performance predictions than when considering constant properties. The effective thermal conductivity will be based on the experimental data obtained by [23].

For the two phase mode with constant properties the thermophysical properties of Al_2O_3 are:

$$\rho_p = 3880 \ \frac{kg}{m^3} \ ; \ Cp_p = 773 \ \frac{J}{kg\ K} \ ; \ k_p = 36 \ \frac{W}{m\ K}$$

while the properties of the base fluid are:

$$\rho_{bf} = 998.2\ \frac{kg}{m^3}\ ;\ Cp_{bf} = 4182\ \frac{J}{kg\ K}$$

$$k_{bf} = 5.97\ \frac{W}{m\ K}\ ;\ \mu_{bf} = 9.93\times 10^{-4}\ \frac{kg}{m\ s}$$

3 Numerical method and validation

In order to ensure the accuracy as well as the consistency of numerical results, several non-uniform grids have been submitted to an extensive testing procedure for each of the cases considered.

Three different grids have been tested (20 x 24 x 800, 10 x 12 x 400 and 5 x 6 x 200) in order to get the best trade-off between solution velocity and accuracy.

Results as obtained for a particular test case have shown that for the tube flow problem under consideration, the 10 x 12 x 400 non-uniform grid appears to be satisfactory to ensure the precision of numerical results as well as their independency with respect to the number of nodes used. Such grid has, respectively, 10, 12 and 400 nodes along the radial, tangential (for h covering 0–180°) and axial directions, with highly packed grid points in the vicinity of the tube wall and especially in the entrance region. Different cases have been run for different Re (250, 550 and 1050) and various D/L ratios (1/50, 1/100 and 1/200).

Moreover an integration time step optimizations has been carried out, particularly three different values (0.2, 0.1 and 0.05s) have been tested and at the end the time step of 0.1s is chosen because it represents the proper compromise between solution velocity and accuracy.

The computer model has been successfully validated with correlation reported in [25] for thermally and hydraulically developing flow. The CFD commercial code Fluent was employed to solve the problem by means of finite volume method.

4 Results

Results were carried out employing the single phase and discrete phase models for φ=1 and 4%, Re=1050 and q=10000 W/m^2 either with constant or temperature-dependent properties. In all cases the size of the spherical particles is considered equal to 100 nm.

The simulation has been performed in transient regime simulating 120s in order to reach the quasi steady state condition, with an integration time step of 0.1s.

In fig. 2 it is shown the average wall temperature behaviour in the time for all the models considered with constant properties (Fig. 2a) and variable properties (Fig. 2b) It is possible to observe that increasing the nanoparticles concentration there is a faster transition to the steady state. In the case of constant properties it is shown that there is a slightly difference between the single phase and the discrete model, while in the variable properties model this difference is

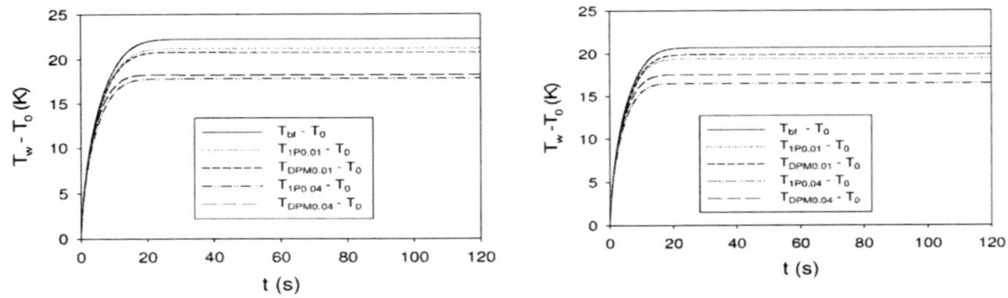

(a) (b)

Figure 2: Average wall temperature: (a) constant and (b) variable properties.

amplified, particularly at the increasing of nanoparticles concentration. The average wall temperature reached by the constant properties model are slightly higher than the ones reached in the variable properties model.

In fig. 3 it is presented the ratio between the average surface heat transfer coefficient in the presence of nanofluid and the one of the basic fluid (water), both in the cases of constant (a) and variables (b) properties. It is possible to observe that both in Fig 3 (a) and (b) for φ=1% there is an increment of the surface heat transfer coefficient around 5%, while for φ=4% there is an increase in the heat transfer coefficient around 20%.

So it is possible to notice how strong is the heat transfer enhancement due to nanoparticles effect, also at relative small concentrations.

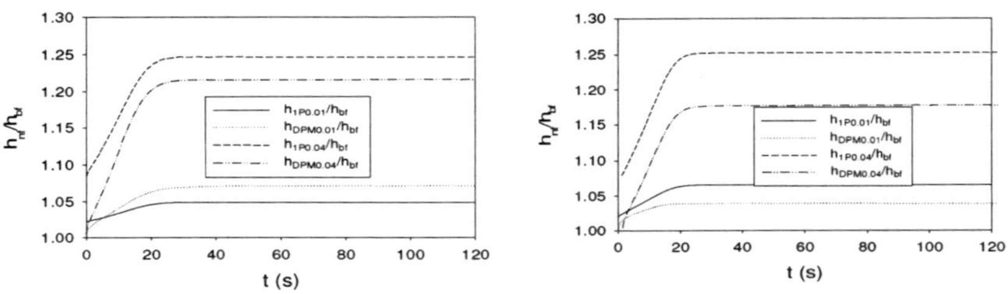

(a) (b)

Figure 3: Surface heat transfer coefficient ratio: (a) constant and (b) variable properties.

In Fig 4 it is presented the temeperature profile along the wall for the cases with constant (a) and variable (b) properties after 120s.

As for the case of constant properties, (Fig. 4a), it is observed that there is a very slight difference in the wall temperature for φ=1% for the one phase and discrete phase model, instead for φ=4% this difference is more evident, particularly the one phase model leads to lower wall temperature.

In the case of variable properties (Fig. 4b), it is shown an appreciable differences between one phase and discrete model for both φ=1% and 4%.

In Fig. 5 it is shown the ratio between the local surface heat transfer coefficient with nanofluids and without nanofluids, the base fluid, along the longitudinal axes after 120s for the two considered cases (constant (a) and variable (b) properties). It is shown that near the inlet section, in the case of the discrete phase model for φ=1% and 4% and for constant and variable properties, a peak which converge to a nearly constant value at z/L=0.4. In the case of single phase model, the ratio, after a slight overshoot in the inlet part (z/L<0.1), tends to grow in a linear way until the outlet section. The discrete phase and single phase almost present the same value for φ=1%, z/L>0.5 and constant properties (Fig. 5 a), while in the other cases there are more marked differences between the two model, which are around the 10%; particularly the single phase model presents higher values.

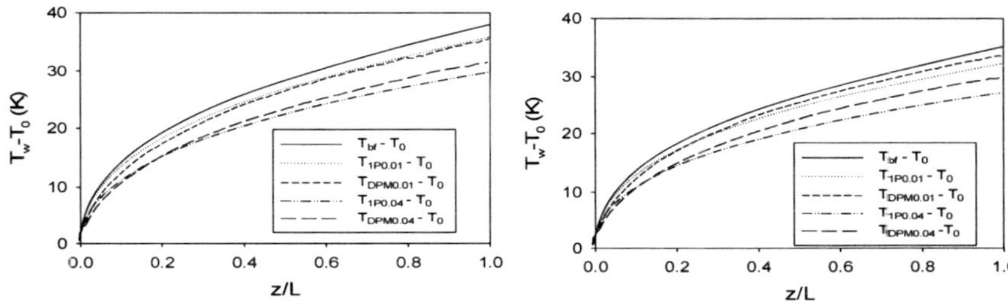

(a) (b)

Figure 4: Wall temperature profile at t=120s: (a) constant and (b) variable properties.

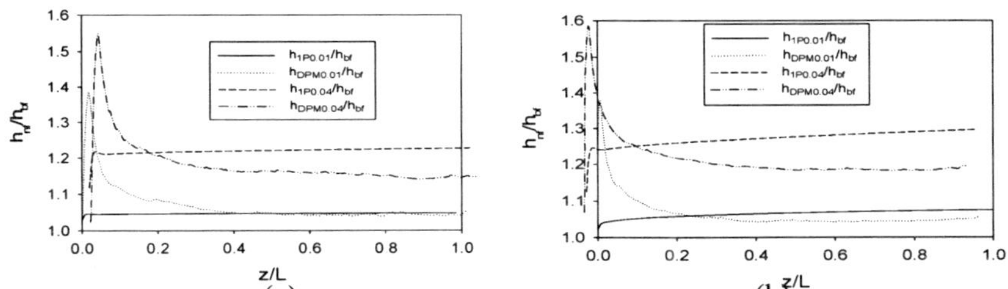

(a) (b)

Figure 5: Local surface heat transfer coefficient after 120s: constant (a) and variable (b) properties.

In Fig. 6 the axial velocity profiles, for z/L=0.5 and t=120s, are presented for the constant (a) and variable (b) properties cases. In both cases the highest velocity values are detected for φ=4% and in the case of variable properties the single and discrete phase models give, practically, the same results, whereas in the case of constant properties there are slight differences only for $r/r_0<0.5$. Also for φ=1% the single and discrete phase profiles are very close.

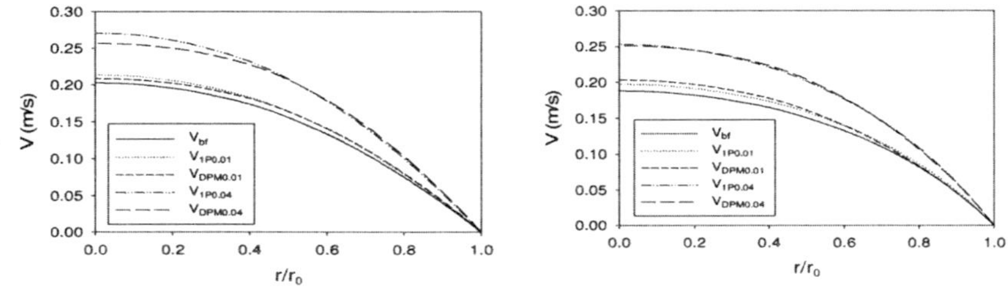

(a) (b)

Figure 6: Axial velocity profile at z/L=0.5 and t=120s.

Table 2: Shear stress ratios comparison.

	Constant Properties		**Variable Properties**	
	1P	DPM	1P	DPM
τ/τ_{bf} (φ=1%)	1.14	1.16	1.18	1.12
τ/τ_{bf} (φ=4%)	1.99	1.54	1.97	1.52

In table 1, the ratio between average shear stress after 120s in presence of nanoparticles and the one of the base fluid is shown. It is possible to observe how it increases with the nanoparticles volume concentration.

5 Conclusions

In this paper the fluid-dynamic and thermal behaviours of water-Al_2O_3 nanofluids flowing inside a tube heated at uniform heat flux were numerically investigated in transient conditions and for laminar flow. Four models were employed: single and two-phase models with either constant or temperature-dependent properties. Results clearly showed that the inclusion of nanoparticles produced a considerable increase of the heat transfer with respect to that of the base liquid. Heat transfer enhancement was increasing with the particle volume concentration. However it was accompanied by increasing wall shear stress values, so it is important to evaluate the trade-off between heat transfer enhancement and shear stress increase.

Nomenclature

Cp	Specific Heat of the fluid, J/kgK	r	Radial coordinate, m
D	Tube diameter, m	r_0	Tube radius, m
d	Particles diameter, m	Re	Reynolds number, $Re=V_0D/\mu$
h	Heat transfer coefficient ($h=q/(T-T_0)$ W/m^{2}K	t	Time, s
k	Thermal conductivity of the fluid W/mK	T	Fluid temperature, K

L	Tube length, m	V	Axial velocity, m/s
P	Pressure, Pa	z	Axial coordinate, m
q	Wall heat flux, W/m^2		

Greek Symbols

φ	Particle volume concentration	ρ	Fluid density, kg/m^3
μ	Fluid dynamic viscosity, kg/ms	τ	Wall shear stress, Pa

Subscripts

av	average value	p	Refers to particle property
bf	Refers to base-fluif	w	Value at wall
nf	Refers to nanofluid property	0	Refers to reference condition
DPM	Discrete Phase Model	1P	Mono Phase

Acknowledgements

This work was supported by SUN with a 2007 grant and MUR with the "EliosLab" program.

References

[1] Maxwell, J.C., A Treatise on Electricity and Magnetism, second ed., Oxford University Press, Cambridge, 1881

[2] Eastman, J.A., Phillpot, S.R., Choi, S.U.S., & Keblinski, P., Thermal transport in nanofluids, *Annu. Rev. Mater. Res.*, 34, pp.219–46, 2004.

[3] Keblinski, P. Eastman, J.A., & Cahill, D.G., Nanofluids for thermal transport, *Materials Today*, **8 (6)**, pp. 36–44, 2005.

[4] Das, S. K., Choi, S. U. S., & Patel, H. E., Heat transfer in nanofluids-a review, *Heat Transfer Engineering*, **27 (10)**, pp. 3–19, 2006.

[5] Buongiorno, J., Convective transport in nanofluids, *ASME J. Heat Transfer*, **128**, pp. 240–250, 2006.

[6] Wang, X.-Q., & Mujumdar, A. S., Heat transfer characteristics of nanofluids: a review, *Int. J. Thermal Sciences*, **46**, pp. 1–19, 2007.

[7] Keblinski, P., Phillpot, S.R., Choi, S.U.S., & Eastman, J.A., Mechanisms of heat flow in suspensions of nano-sized particles (nanofluid), *Int. J. Heat Mass Transfer*, **45**, pp. 855–863, 2002.

[8] Pak, B.C., & Cho, Y.I., Hydrodynamic and heat transfer study of dispersed fluids with submicron metallic oxide particles, *Exp. Heat Transfer*, **11**, pp. 151–170, 1998.

[9] Xuan, Y.M., & Li, Q., Investigation on convective heat transfer and flow features of nanofluids, *J. Heat Transfer* **125**, 151–155, 2003.

[10] Wen, D., & Ding, Y., Experimental investigation into convective heat transfer of nanofluids at the entrance region under laminar flow conditions, *Int. J. Heat Mass Transfer*, **47 (24)**, pp. 5181–5188, 2004.

[11] Maiga, S.E.B., Nguyen, C.T., Galanis, N., & Roy, G., Heat transfer behaviours of nanofluids in a uniformly heated tube, *Superlattices Microstruct.*, **35**, pp. 543–557, 2004.

[12] Maiga, S.E.B., Palm, S.J., Nguyen, C.T., Roy, G., & Galanis, N., Heat transfer enhancement by using nanofluids in forced convection flows, *Int. J. Heat Fluid Flow*, **26 (4)**, pp. 530–546, 2005.

[13] Palm, S.J., Roy, G., & Nguyen, C.T., Heat transfer enhancement with the use of nanofluids in radial flow cooling systems considering temperature dependent properties, *Appl. Thermal Eng.*, **26**, pp. 2209–2218, 2006.

[14] Akbarinia, A., & Behzadmehr, A., Numerical study of laminar mixed convection of a nanofluid in horizontal curved tubes, *Appl. Thermal Eng.*, **27**, pp. 1327–1337, 2007.

[15] Behzadmehr, A., Saffar-Avval, M., & Galanis, N., Prediction of turbulent forced convection of a nanofluid in a tube with uniform heat flux using a two phase approach, *Int. J. Heat Fluid Flow*, **28**, pp. 211–219, 2007.

[16] Mirmasoumi, S., & Behzadmehr, A., Numerical study of laminar mixed convection of a nanofluid in a horizontal tube using two-phase mixture model, *Applied Thermal Engineering*, **28**, 717–727, 2008.

[17] Chiacchio, F., Manca, O., & Nardini, S., Numerical investigation of single phase forced convection of nanofluids in circular tubes, *Proc. 2007 ASME-JSME Thermal Engineering Conference and Summer Heat Transfer Conference*, paper HT2007-32855, July 8-12, 2007, Vancouver, BC, Canada, 2007.

[18] Fluent Incorporated, *Fluent 6.2*, *User Manual*, 2006.

[19] Ounis, H., Ahmadi, G., & McLaughlin, J. B., Brownian diffusion of submicrometer particles in the viscous sublayer, J. *Colloid Interface Science*, **143 (1)**, pp. 266–277, 1991.

[20] Masuda, H., Ebata, A., Teramae, K., & Hishinuma, N., Alteration of thermal conductivity and viscosity of liquid by dispersing ultra-fine particles (dispersion of γ -Al2O3, SiO2 and TiO2 ultra-fine particles), *Netsu Bussei* (in Japanese), **4 (4)**, pp. 227–233, 1993.

[21] Lee, S., Choi, S.U.-S., Li, S., & Eastman, J.A., Measuring thermal conductivity of fluids containing oxide nanoparticles, *J. Heat Transfer*, **121**, pp. 280–289, 1999.

[22] Wang, X., Xu, X., & Choi, S.U.-S., Thermal conductivity of nanoparticles–fluid mixture, *J. Thermophys. Heat Transfer*, **13 (4)**, pp. 474–480, 1999.

[23] Putra, N., Roetzel, W., & Das, S.K., Natural convection of nanofluids, *Heat Mass Transfer*, **39**, pp. 775-784, 2003.

[24] Hamilton, R.L., & Crosser, O.K., Thermal conductivity of heterogeneous two-component systems, *I & EC Fundamentals*, **1 (3)**, pp. 187–191, 1962.

[25] Churchill, S.W., & Ozoe, H., Correlations for forced convection with uniform heating in flow over a plate and in developing and fully developed flow in a tube, *J. Heat Transfer*, **95**, pp. 78-84, 1973.

Numerical investigation of natural convection of air in vertical divergent channels

O. Manca, S. Nardini, D. Ricci & S. Tamburrino
Dipartimento di Ingegneria Aerospaziale e Meccanica,
Seconda Università degli studi di Napoli, Italy

Abstract

A numerical transient analysis of natural convection in air in vertical divergent channels is accomplished. The channel walls are heated at uniform heat flux, the problem is two-dimensional and laminar and the full Navier-Stokes and energy equations are employed. Results in terms of average wall temperature profile, as a function of the time, wall temperature distributions for inclination angle and minimum wall spacing are presented. The simulation allows the detection of complex structures of the flow inside and outside the channel. Temperature profiles as a function of time show overshoot and undershoot. A comparison between numerical flow patterns and experimental flow visualizations are also given and a good agreement is observed.
Keywords: natural convection, divergent channel, transient analysis.

1 Introduction

Natural convection in channels and parallel plates received very much attention for its application in engineering as reviewed in [1, 2]. These configurations have been employed in chemical vapor deposition reactors, solar collectors, nuclear reactors, heat exchangers and thermal control of electronic systems [3,4]. Nowadays, trends in natural convection heat transfer are oriented toward either the seeking of new configurations to enhance the heat transfer parameters or the optimization of standard configurations [5, 6]. In fact, recent research efforts have been devoted to define simple geometrical modification of the channel configurations to improve its thermal performance in natural convection [5–7]. Among them divergent vertical channels with two heated principal walls are very simple ones [9–12].

WIT Transactions on Engineering Sciences, Vol 61, © 2008 WIT Press
www.witpress.com, ISSN 1743-3533 (on-line)
doi:10.2495/HT080021

The divergent or convergent channels, even if symmetrically heated could have a significant view factor toward the surrounding which depends on the aperture angle. Then it is interesting to investigate the radiative effects on thermal performance of natural convection in air in divergent or convergent channels. About the divergent channel Agonafer and Watkins [8] studied numerically one-dimensional natural convection in a divergent channel for inclination angle values up to 8 degrees. An in-depth experimental study was carried out by Sparrow and Ruiz [9]. The principal walls were at uniform temperature and experiments were performed with water. Flow visualization indicated the presence of boundary layers adjacent to the channel walls with a recirculation loop occupying the remainder of the cross section. A single correlation between channel Nusselt and Rayleigh numbers for convergent, divergent and parallel walled channels was proposed.

Recently, Bianco et al. [10] carried out an experimental investigation on air natural convection in divergent channels with uniform heat flux at the principal walls. They found that the larger the Rayleigh number the worst the thermal performance of the channel, for several divergence angles larger than zero degrees and for low Rayleigh numbers. Flow visualization of air natural convection in a vertical divergent channel symmetrically heated at the walls, with a uniform heat flux, for different values of the divergence angle, the minimum channel spacing and the dissipated heat flux was carried in [11]. Results showed that the flow was laminar when the divergence angle was small (θ=2°) and that the larger the divergence angle the larger the penetration depth of ambient air in the channel from its upper end section, which forms an air downflow in the central region of the channel and an air upflow parallel to the boundary layer.

A numerical simulation for vertical diverging and converging channels, with laminar steady state natural convection, was performed in [12]. The heated length of the walls was lower or equal to the wall length and it was isothermal. The adiabatic zone of the walls was placed either at the bottom end of the channel or at the top end.

Marcondes et al. [13] carried out a numerical investigation on natural convection in parallel, convergent and divergent channels using a fully elliptic procedure considering only the simple channel as computational domain. Both the channel walls were at uniform temperature and results were given for Prandtl number ranging from 0.7 to 88. A correlation for average Nusselt number, Rayleigh number, in terms of maximum channel width and channel aspect ratio, and Prandtl number was proposed. They found that for convergent channels a recirculation region in the outlet zone was observed.

The analysis of wall temperature profiles presented in [11] and the observations provided in [12] suggests that a deeper insight into the fluid flow in natural convection in a vertical divergent channel should be given. Average wall temperature profiles depending on time, temperature distributions along the duct walls, and comparisons between flow visualization for experimental cases and stream function contours given for numerical simulations are presented in this paper for air natural convection in a vertical divergent channel symmetrically heated at the walls, for different divergence angles and channel spacing.

2 Mathematical description and governing equations

The physical domain under investigation is shown in Fig. 1. It consists of two non-parallel plates that form a vertical divergent channel. Both plates are not thermally conductive and heated at uniform heat flux set equal to 120 W/m^2. The imbalance between the temperature of the ambient air, T_o, and the temperature of the heated plates draws an air flow rate into the channel. The transient flow in the channel is assumed two-dimensional, laminar, incompressible, with negligible viscous dissipation. All thermophysical properties of the fluid are assumed constant, except for the dependence of density on the temperature (Boussinesq approximation) which gives rise to the buoyancy forces. The thermophysical properties of the fluid are evaluated at the ambient temperature, T_o, which is assumed to be 300 K in all cases. With the above assumptions, the governing equations in primitive variables are:

$$\frac{\partial u}{\partial x}+\frac{\partial v}{\partial y}=0 \tag{1}$$

$$\frac{\partial u}{\partial t}+u\frac{\partial u}{\partial x}+v\frac{\partial u}{\partial y}=-\frac{1}{\rho}\frac{\partial p}{\partial x}+\nu\left(\frac{\partial^2 u}{\partial x^2}+\frac{\partial^2 u}{\partial y^2}\right) \tag{2}$$

$$\frac{\partial v}{\partial t}+u\frac{\partial v}{\partial x}+v\frac{\partial v}{\partial y}=-\frac{1}{\rho}\frac{\partial p}{\partial y}+\nu\left(\frac{\partial^2 v}{\partial x^2}+\frac{\partial^2 v}{\partial y^2}\right)-g\beta\left(T-T_o\right) \tag{3}$$

$$\frac{\partial T}{\partial t}+u\frac{\partial T}{\partial x}+v\frac{\partial T}{\partial y}=a\left(\frac{\partial^2 T}{\partial x^2}+\frac{\partial^2 T}{\partial y^2}\right) \tag{4}$$

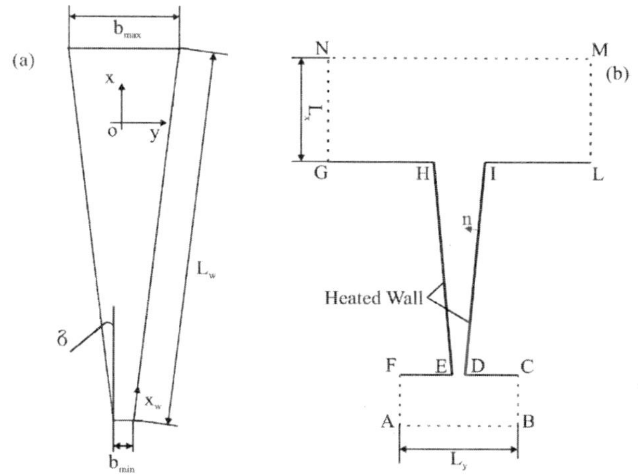

Figure 1: Geometrical configuration: (a) physical domain; (b) computational domain.

The channel modified Rayleigh number and average Nusselt numbers are considered:

$$Ra_b = \frac{g\beta\, q_w b^5}{\nu^2 kL} Pr \tag{5}$$

$$Nu_b = \frac{q_w b}{\left(T_{w,av} - T_o\right)k} \tag{6}$$

where b is b_{min} or b_{av} or b_{max}

$$T_{w,av} = \frac{1}{L}\int_0^L T_w(x)dx\,,\; q_w = \frac{1}{L}\int_0^L q_w(x)dx \tag{7}$$

3 Numerical procedure

Since the two plates are placed in an infinite medium, from a numerical point of view the problem is solved with reference to a computational domain of finite extent, as depicted in figure 1b, and by following the approach given in [12, 14]. A finite extension computational domain is employed to simulate the free-stream condition and allows to account for the diffusive effects, peculiar in the elliptic model. The boundary conditions are given in Table 1. The numerical model is solved using the commercial code FLUENT [15]. The segregated solution method is chosen to solve the governing equations, which are linearized implicitly with respect to the dependent variable equation. The second-order upwind scheme is chosen for the unsteady energy and momentum equations [15]. The SIMPLE scheme is chosen to couple pressure and velocity. Computation starts with zero values of velocities and with pressure and temperature values equal to the ambient ones, initial condition of the problem. The convergence criteria of 10^{-6} for the residuals of velocity components and of 10^{-8} for the residuals of the energy are assumed.

A grid dependence test is accomplished to realize the more convenient grid size by monitoring variables like average wall temperature, average Nusselt number and average outlet velocity for a divergent channel system with a spacing b = 10, divergent angle equal to 2° and $Ra_{bmin} = 3.3x10^3$.

About the time step a value of 0.1 s is employed in the investigation after some tests on three different time steps (0.05, 0.1 and 0.2 s). Three mesh grids are analyzed, by doubling the number of nodes (from 27000 to 112000). It is observed that by doubling the number of the nodes in the channel, a variation of about 0.2% in terms of average wall temperature, 0.5% in terms of bulk temperature at the outlet section and 0.2% in terms of average velocity at the outlet section are observed. The mesh size with 56x560 nodes is employed in this investigation because it ensures a good compromise between the computational time and the accuracy requirements. The reservoir at the bottom have horizontal (L_y) and vertical (L_x) lengths set equal to twenty times the minimum spacing distance between the walls while the for the reservoir on the top L_y and L_x are, respectively, thirty-five and forty times than the minimum spacing.

The validation of results is accomplished by comparing a numerical simulation with the experimental results. According to Bianco et al. [10,11] the conductive and radiative heat losses are about 14% of the Ohmic wall heat flux. So a simulation for $\theta = 2°$ and $b_{min} = 10$ mm has been carried out by applying a constant heat flux equal to 103 W/m^2 and a maximum difference of 2 K has been detected by comparing numerical results with experimental data.

Table 1: Boundary conditions for the fluid domain.

Wall	u	v	T	Wall	u	v	T
AF and BC	$\frac{\partial u}{\partial y}=0$	$\frac{\partial v}{\partial y}=0$	$T=T_o$	EH	u=0	v=0	$k_f \frac{\partial T}{\partial n} = -q_w$
AB	$\frac{\partial u}{\partial x}=0$	$\frac{\partial v}{\partial x}=0$	$T=T_o$	MN	$\frac{\partial u}{\partial x}=0$	$\frac{\partial v}{\partial x}=0$	$\begin{cases}\frac{\partial T}{\partial x}=0 & \text{if } u>0\\ T=T_o & \text{if } u<0\end{cases}$
EF, CD, GH, IL	u=0	v=0	$\frac{\partial T}{\partial x}=0$	GN	$\frac{\partial u}{\partial y}=0$	$\frac{\partial v}{\partial y}=0$	$\begin{cases}\frac{\partial T}{\partial y}=0 & \text{if } v<0\\ T=T_o & \text{if } v>0\end{cases}$
DI	u=0	v=0	$k_f \frac{\partial T}{\partial n} = q_w$	LM	$\frac{\partial u}{\partial y}=0$	$\frac{\partial v}{\partial y}=0$	$\begin{cases}\frac{\partial T}{\partial y}=0 & \text{if } v>0\\ T=T_o & \text{if } v<0\end{cases}$

4 Results and discussions

Average wall temperature profiles as a function of time, temperature profiles along the channel walls, and comparisons between flow patterns for experimental cases and stream function contours given for numerical simulations are presented for different divergence angles and channel spacing. This is useful in order to remark the thermal and fluid dynamic behaviours of natural convection in vertical divergent channel.

Fig. 2 reports the average wall temperature profiles for different angles and minimum wall spacings. It is observed, for $\theta = 0°$, a reduction of wall temperature by increasing the channel spacing up to $b_{min} = 20$ mm. For $b_{min} = 7.0$ mm (Fig. 2a) maximum average wall temperature is detected for $\theta = 0°$ while for $b_{min} = 10$, 20 and 40 mm the simulations for $\theta = 5°$ reveal the maximum values at steady state. However, for $\theta = 10°$ and 15° the steady state condition is not reached and oscillations are present, in accordance with [12]. The amplitude of the oscillations increase increasing the spacing between the walls. It is interesting that up to $b_{min} \leq 10$ mm, the minimum average wall temperature are attained for $\theta = 10°$ and 15° whereas for $b_{min} \geq 20$ mm the minimum value is obtained for $\theta = 0°$. For all divergence angles temperature overshoots are detected, the temperature profiles show that there is an initial increase of temperature up to the maximum value. The overshoots present different trends depending on the divergence angles. In fact, for $\theta = 0°$, $2°$ and $5°$ the overshoot is attained at a lower instant at about 2 s whereas for $\theta = 10°$ and $15°$ it is reached for time not less than 10 s and this time increases increasing the wall spacing b_{min}.

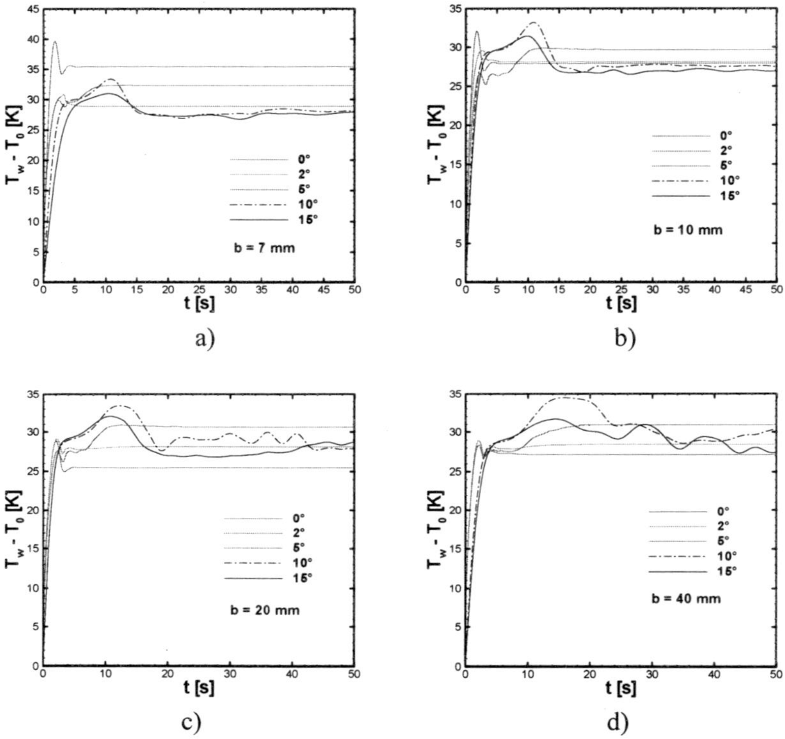

Figure 2: Average wall temperature profiles depending on time: a) b_{min} = 7.0 mm; b) b_{min} = 10 mm; c) b_{min} = 20 mm and d) b_{min} = 40 mm.

Fig. 3 shows the wall temperature distributions at t = 50 s. Maximum wall temperatures decrease for greater wall spacings. For b_{min}=7 mm, in Fig. 3a, the highest wall temperature is noted for the parallel plate configuration, $\theta = 0°$. The difference between the maximum wall temperature for $\theta = 0°$ and $\theta = 10°$ is about 15 °C. The configuration with a divergence angle equal to 2° presents the minimum value of maximum wall temperature, though the differences respect to the values for the other divergence angles are very small. For b_{min}=10 mm, in Fig. 3b, the largest maximum wall temperature is again attained for $\theta = 0°$ but the difference respect to the values for the other configurations is significantly decreased, it is about 3 °C. It is interesting to observe that for all considered b_{min} values the lowest maximum wall temperature for $\theta = 0°$ is reached at

b_{min}=40 mm, in Fig. 3d. However, for the considered configurations it is convenient to give a slight divergence to the channel for lower $b_{min} \leq 10$ mm.

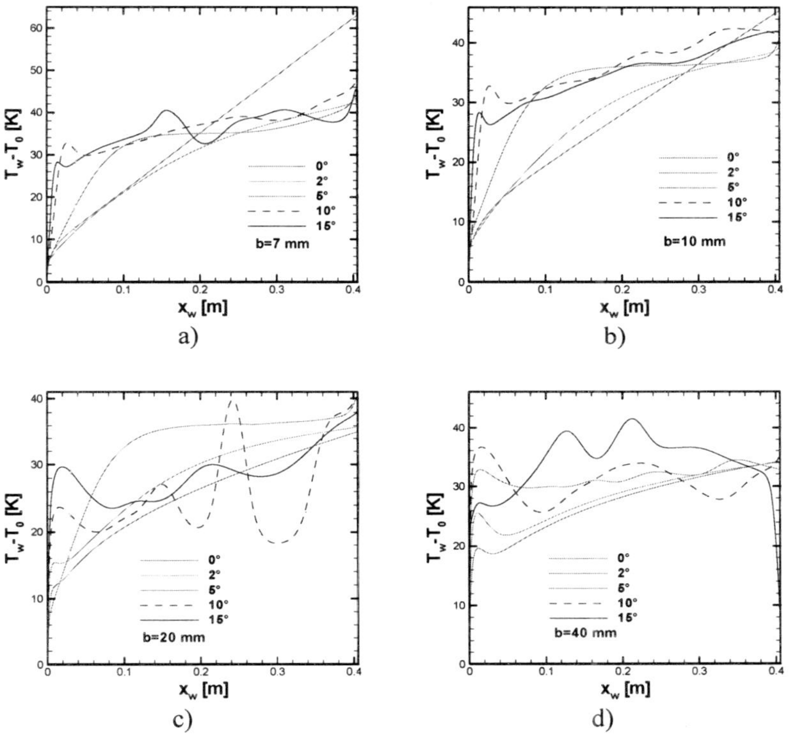

Figure 3: Temperature distributions along the channel walls at t = 50 s: a) b_{min} = 7 mm; b) b_{min} = 10 mm; c) b_{min} = 20 mm and d) b_{min} = 40 mm.

Fig. 4 shows a comparison between flow patterns obtained for experimental cases and stream function contours of the numerical simulations for θ = 2°, 5° and 10° and b_{min} = 20 mm. In Fig. 4a and 4d one can remark the air laminar flow close to the wall, very similar to one into a parallel channel. Fig. 4b and 4e show that increasing the divergence angle the air layer adjacent to the wall in the lower half height of the channel thickens. At mid-height the air layer becomes thinner because of the parallel upflow of the air towards the channel exit, which occurs together with an ambient air downflow from the exit section. Considering θ = 10° a deeper penetration of the ambient air and an upflow closer to the boundary layer which improve heat transfer rate are revealed. At x values larger than about 100 mm the flow is very wavy because of the interaction between the air close to the wall and the vertical air downflow in the central region of the channel.

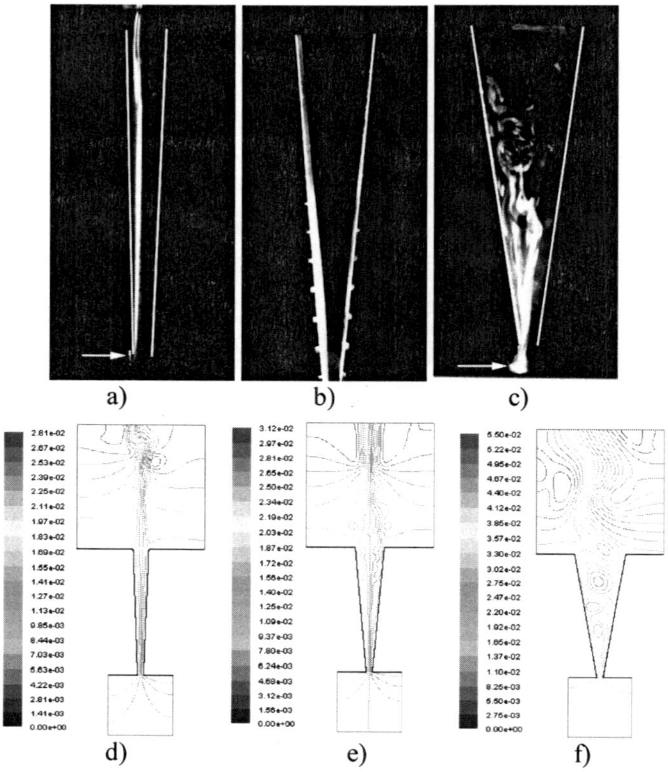

Figure 4: Flow patterns (a, b, c) of experimental cases and stream function contours (d, e, f) of numerical simulations for b_{min} = 20 and $\theta = 2°$, 5° and 10° at t = 50s.

5 Conclusions

Transient natural convection in a divergent vertical channel, with the wall heated symmetrically at uniform heat flux, was numerically investigated. A two dimensional transient laminar regime was considered and solved by a finite volume method employing the commercial code Fluent. Results showed that the average wall temperature as a function of time, for divergence angles greater than 5° presented oscillations which increased their amplitude at increasing the wall spacing. In all configurations overshoot and undershoot were detected. Temperature profiles along the walls showed that for angles greater than 5° periodic trends due to the cold inflow in the divergent channel were detected. In fact, the flow visualization and the stream function fields confirmed the disturbances inside the channel for higher divergence angles, ≥ 10°. For lower channel spacing, 7 mm, a small divergence angle allows to reduce significantly the maximum wall temperature whereas for higher wall spacings the divergence of the channel do not provide a decrease in terms of maximum wall temperature. The comparison between experimental flow visualizations and stream function fields obtained numerically were in very good agreement. Moreover, the complex flow structures detected in the channel allow the configuration with a divergence angle equal to 5° as a critical configuration to be defined.

6 Nomenclature

b	channel spacing, m
g	acceleration of gravity, m s^{-2}
Gr	channel Grashof number
k	thermal conductivity, $Wm^{-1} K^{-1}$
L	channel length, m
Nu	Nusselt number
Pr	Prandtl number
q	heat flux, W m^{-2}
Ra	channel Rayleigh number
T	temperature, K
x, y	coordinates along the plate, m

Greek symbols

β	volumetric coefficient of expansion, K^{-1}
θ	divergence angle, deg
ν	kinematic viscosity, $m^2 s^{-1}$

Subscripts

av	average
f	fluid
min	minimum
max	maximum
o	ambient
w	wall

Acknowledgement

This work was supported by SUN with a 2007 grant.

References

[1] Manca, O., Morrone, B., Nardini, S., & Naso, V., Natural Convection in Open Channels in Computational Analysis of Convection Heat Transfer, Sundén B. and Comini G. Editors, WIT Press, Southampton, Chap. 7, pp. 235–278, 2000.

[2] Bianco, N., Morrone, B., Nardini, S., & Naso, V., Air natural convection between inclined parallel plates with uniform heat flux at the walls, *International Journal of Heat and Technology*, **18**, pp. 23–46, 2000.

[3] Gebhart, B., Jaluria, Y., Mahajan, R. & Sammakia, B., *Buoyancy-Induced Flows and Transport*, Hemisphere Publ. Corp., Washington, D.C, 1998.

[4] Raithby, G. D. & Hollands, K. G. T., Natural Convection, in *Handbook of Heat Transfer*, Rohsenow, W. M., Hartnett, J. P., Cho, Y. I. editors, McGraw-Hill, New York, Chap. 7, pp. 4.1–4.99, 1998.

[5] Bejan, A., *Shape and Structure, from Engineering to Nature*, Cambridge University Press, Cambridge, 2000.

[6] Ledezma, G. A., & Bejan, A., Optimal geometric arrangement of staggered vertical plates in natural convection, *ASME Journal of Heat Transfer*, **119**, pp. 700–708, 1997.

[7] Kim, S. J., & Lee, S. W., *Air Cooling Technology for Electronic Equipment*, CRC Press, Boca Raton, FL, 1998.

[8] Agonafer, D., & Watkins, C. B., Numerical solution of natural convection between diverging plates, *ASME Paper 84-WA*, **HT 32**, 1983.

[9] Sparrow, E. M., & Ruiz, R., Experiments on natural convection in divergent vertical channels and correlation of divergent, convergent, and parallel-channel nusselt numbers, *International Journal of Heat and Mass Transfer*, **31**, pp. 2197–2205, 1988.

[10] Bianco, N., Manca, O., Nardini, S. & Naso, V., Air natural convection in a divergent channel with uniformly heated plates, *Proceedings, XIX Congresso Nazionale sulla Trasmissione del Calore UIT*, pp. 89–94, 2001.

[11] Bianco, N., Manca, O., Nardini, S. & Naso, V., Visualization of natural convection in a vertical divergent channel plates, *Proceedings, Eurotherm Seminar 71 on Visualization, imaging and data analysis in convective heat and mass transfer*, pp. 117–122, Reims, October 28–30, 2002.

[12] Bejan, A., da Silva, A. K., & Lorente, S., Maximal heat transfer density in vertical morphing channels with natural convection, *Numerical Heat Transfer Part A*, **45**, pp. 135–152, 2004.

[13] F. Marcondes, V. de Souza Melo, & J. M. Gurgel, Numerical analysis of natural convection in parallel, convergent, and divergent open-ended channels, *International Journal of Numerical Methods for Heat and Fluid Flow*, **16 (3)**, pp. 304–323, 2006.

[14] A. Andreozzi, & O. Manca, Thermal and fluid dynamic behaviour of symmetrically heated vertical channels with auxiliary plate, *International Journal Heat Fluid Flow*, **22**, pp. 424–432, 2001.

[15] Fluent Incorporated, Fluent 6.1, *User's Manual*, 2003.

Section 2
Heat exchangers

Numerical simulation of fluid flow in a monolithic exchanger related to high temperature and high pressure operating conditions

F. Selimovic & B. Sundén
Division of Heat Transfer Faculty of Engineering, LTH Box 118, 221 00 Lund, Sweden

Abstract

The purpose of this work is the enhancement of performance by a new design of monolithic heat exchangers under steady-state operating conditions. Heat transfer phenomena and hydrodynamics have been studied and visualized by computational fluid dynamics (CFD). Further, a simple gas distribution system has been analyzed in purpose to find the best performance of the exchanger. To achieve this, 3D simulations of air flow were performed. One of characteristics of monolithic structures is the low pressure drop and high heat transfer coefficient. This is because they operate in the laminar flow regime and have high compactness. However, some experimental studies show that when two fluids are introduced into monolith channels, the manifolds cause severe pressure losses. Therefore, in this work, turbulent flow regime at low Reynolds numbers has been investigated to find the difference and better understanding of these structures, which are of interest in high temperature applications today. The simulation shows that the pressure drop of the gas flow distributor is a key parameter affecting the heat transfer in the exchanger channels.
Keywords: high temperature heat exchangers, gas flow maldistribution, monolithic heat exchangers, computational fluid dynamics.

1 Introduction

Monolith structures are used in industry today and they are produced by extrusion techniques. They are uni-body structures composed of interconnected

WIT Transactions on Engineering Sciences, Vol 61, © 2008 WIT Press
www.witpress.com, ISSN 1743-3533 (on-line)
doi:10.2495/HT080031

repeated cells or channels. Mainly these structures are used in single-fluid applications and are made of metallic materials, e.g., the monolithic exhaust structure in automotive applications, heat sinks, etc. If the two fluids can exchange heat and/or mass there would be greater reaction rates and exchangers with small dimensions would be advantageous. This is the case if these structures contain micro- or mini-channels because the heat and mass transfer coefficients exceed the values of conventional devices. Their vacuum tightness, the high pressure stability of several hundreds bar, and the flame-arresting properties of the fine micro channels, even allow for reactions with high educt concentrations in the explosion range, [1].

Especially two-fluid monoliths made of ceramics with micro-channels would open up new fields of application, e.g., high-temperature reactions. However, only few attempts have been made to manufacture monolithic components from ceramics, e.g., by lamination of ceramic green tapes, [2]. Usually high pressure losses through manifolds have also been a drawback so that the design geometry needs to be adjusted to limit pressure losses to acceptable levels. In other words usage of monolithic structures in two-fluid applications would be possible, if carefully designed distribution headers are assembled and could guarantee against excessive flow maldistribution and avoid unacceptable pressure drops.

In a previous work, [3], one manifold design solution was presented. It has been found both experimentally and computationally that the largest pressure drop occurs in manifolds and distributor plates for a two-fluid exchanger. In this work an idea for a monolithic ceramic heat and/or mass reactor/exchanger will be presented. Starting from the geometrical design, results obtained by computational fluid dynamics (CFD) simulation for the pressure losses, flow maldistribution, heat transfer will be given.

The aim of the present study is to present a computational investigation of a monolithic structure using CFD predictions over a wide range of operating conditions used in high temperature application such as catalyst reactors. Heat transfer phenomena and hydrodynamics are studied and visualized by CFD.

2 Two fluid exchanger model

An idea for a new design of the two-fluid exchanger is illustrated in Figure1. In this study five-channels with ten manifold inlet and outlet stacks for each fluid are considered. From Figure 2b) five-channels repeating unit can be observed. Main inlets and outlets with flow distribution into channels are illustrated in Figure 2a). The exchanger is to be operated in counterflow mode, where heat and mass exchange can take place by linear channel arrangement. The cooling fluid, Fluid 2, enters the exchanger at the bottom manifold at right where cold fluid, Fluid 1 does it at the top, Figure 1.

This solution for the flow manifolds differs from previous work [3] because when linear channel arrangement is introduced, one can avoid having complex header system in contrast to system having a checkerboard channel arrangement. Computer model of the exchanger is limited by computational expenses. For that purpose the computational model of the monolith exchanger has been scaled

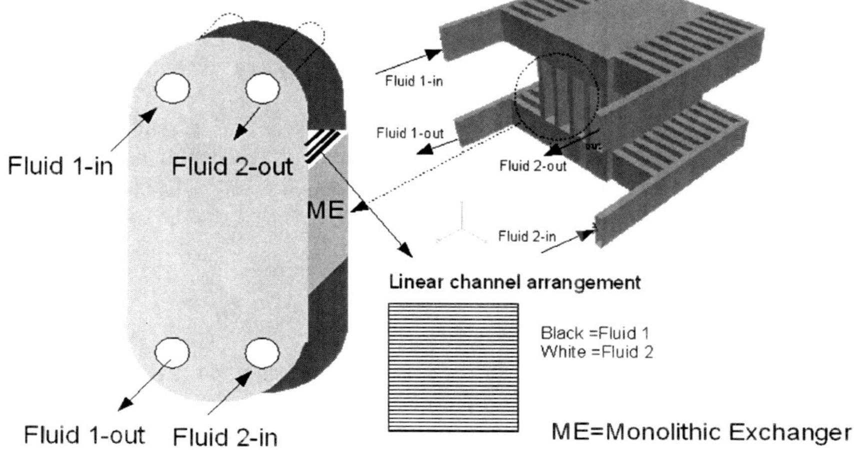

Figure 1: Principle setup of a counterflow mini-channel monolithic heat exchanger.

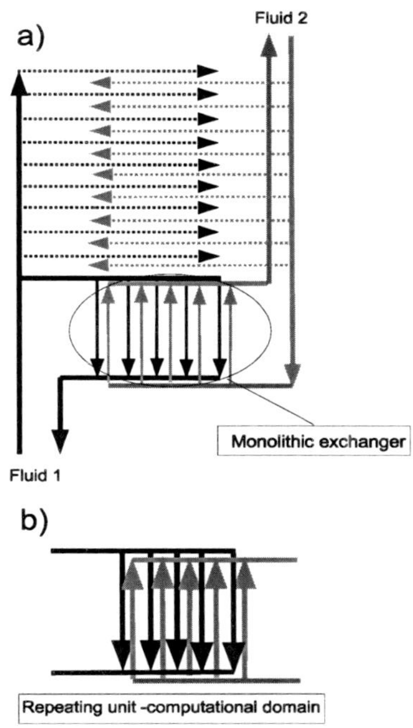

Figure 2: (a) Guided mass flows from inlet manifold into exchanger channels. (b) Computational domain for one repeated unit.

down to centimetre level. The model of the exchanger will be represented only by one "repeating unit" which is illustrated in Figure 3. A repeated unit with inlet and outlet manifolds is 200 mm long and has a cross-section of 10 mm x 10 mm with 5 channels with height of 100 mm. The channels have the same cross-section as the manifolds.

The present model contains 5 monolithic channels but in real systems more channels are possible and should be considered from design point of view. However, heat transfer and pressure drop will be analyzed for this model caused by maldistribution and the following correction equation will cover cases for the 5 channels considered.

3 CFD modelling

The commercially available CFD code FLUENT is used to predict steady state heat transfer performance of the monolithic structure. Figure 3 represents a grid example chosen for the simulations. The working fluid for the cases studied is air. Results are collected for pressure drop, fluid flow maldistribution and heat transfer. Because of earlier experience in experimental analysis of monolith heat exchanger [3], low Reynolds variant of k-ϵ model has been used in simulation to find correction for pressure loss. Therefore, special care is taken to the location of the computational nodes near the wall. Turbulence intensity has been set to 5%.

3.1 Modelling equations and boundary conditions

The equations that govern fluid motion and heat transfer are: the continuity, momentum and energy equations. In modelling of turbulence the standard k-ϵ model with additionally two-layer low Reynolds number model have been used. For further details about the modelling equations, see [4].

A velocity inlet boundary condition (uniform velocity distribution with a direction vector normal to the inlet boundary) was applied at the fluid inlet flow manifold. At the outlet, the boundary condition was assigned as outlet which means that gradients of all flow properties except pressure are set to zero.

If the direction normal to the outlet boundary is denoted by y, the outlet boundary condition is expressed as in [1]. ϕ represents all the scalar variables of interest (temperatures, physical properties etc). The walls are treated as adiabatic besides two channel-walls which are main contributors to the heat transfer in this structure. Pressure drops over the inlet manifold and outlet manifold are calculated by area weighted averages. When the pressures drop across them are known the inlet and outlet boundaries are changed to pressure inlet and pressure outlet which allows using the ideal gas model for density of air [4].

$$\frac{\partial \phi}{\partial y} = 0 \tag{1}$$

The simulations are run for the following cases of Reynolds number occurring in the channels: 30, 150, 500 and 900.

Heat transfer of a rectangular channel has been analyzed for the temperature boundary condition of the third kind: the local wall heat flux is a linear function of the local wall temperature, [5], where both left and right walls of monolithic exchanger channels have been heated at constant wall temperature of 1100K. To obtain a dimensionless representation of the pressure drop due to the manifold and channels, the friction factor f based on adiabatic conditions was introduced according to the Fanning definition:

$$f = \frac{1}{2}\frac{D_h}{\rho U_{bulk}^2}\frac{dP}{dy} \tag{2}$$

The variable local and average Nusselt numbers for the channel left or right wall are based on the convective heat transfer coefficient h_x and it is defined as:

$$Nu = \frac{h\,D_h}{k} = \frac{q_w\,D_h}{k(T_w - T_{bulk})} \tag{3}$$

$$\overline{Nu} = \frac{\bar{h}D_h}{k} = \frac{q_w D_h}{k(\overline{T_w} - T_{bulk})} \tag{4}$$

Where T_{bulk} is the bulk stream-wise flow temperature in the cross section of the duct.

$$T_{bulk} = \frac{\int T|U|dA}{\int |U|dA} \tag{5}$$

The dimensionless axial distance y^* in the flow direction for the hydrodynamic entrance region is defined as

$$y^* = \frac{y}{D_h Re} \tag{6}$$

Friction factor has been compared with the cases for fully developed laminar flow and for the case of monolithic exchanger with checkerboard channel arrangement, found in [3]. For laminar case, Fanning friction factor is correlated to [6]:

$$fRe = 14.227 \tag{7}$$

and for the case of checkerboard arrangement :

$$\mathrm{r}f = 110.5\,Re^{-0.2615} \tag{8}$$

3.2 Computational grid and solution procedure

A structured grid is used for the simulations in the computational domain.

At the end of iterations, the maximum changes in the value of each dependent variable are calculated. The criterion of the maximum changes is set to 1E-4. To achieve grid independence several sensitivity grid-tests were carried out. Test cases were performed for grids with 135000, 200000, 300000, and 600000 computational elements. Standard k-ϵ model does not give acceptable grid independence even if the amount of computational cell is increased up to 600000, Figure 6. Therefore, the standard k-ϵ model turns out to be inadequate to capture all the small-scale feature of turbulence (without using unsteady state equations). A necessity of applying another turbulence model is obvious.

An enhanced wall treatment is used in order to find better convergence. Enhanced wall treatment is a near-wall modelling method that combines a two-layer model with enhanced wall functions. As can be observed in Figure 6 (left), the total pressure profile calculated with enhanced wall treatment model with

135000 cells matches the result achieved by the standard k-ϵ model with 600 000 cells. Figure 6 (right), shows influence of different grid sizes using enhanced wall treatment model. Convergence has been achieved with 200 000 computational elements. Therefore, this grid size is used for the rest of simulations.

Numerical solution has been performed by means of the commercial software FLUENT 6.3.26 where each of differential equations has been transformed into the general transport equation form:

$$div(\rho\phi u) = div(\Gamma\ grad\ \phi) + S_\phi \quad (9)$$

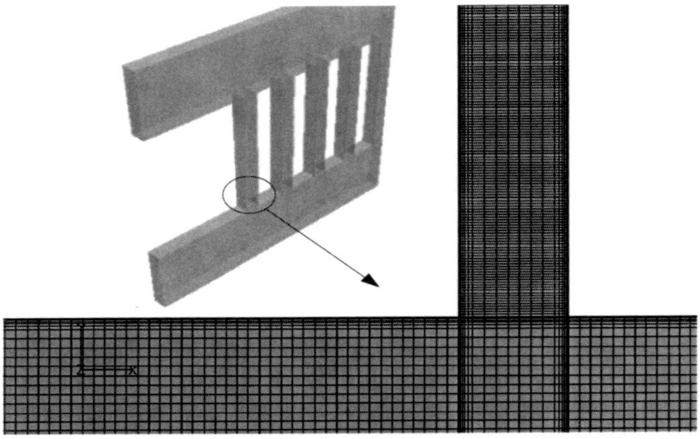

Figure 3: The computational domain and detailed view of grid used in calculations. Finer mesh is used near walls.

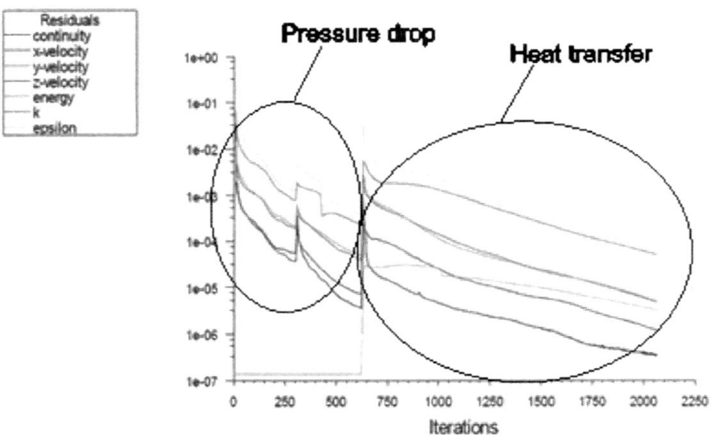

Figure 4: Iterations procedure for each steady-state simulation. (Observe residual jump when QUICK method is turned on in pressure drop simulations).

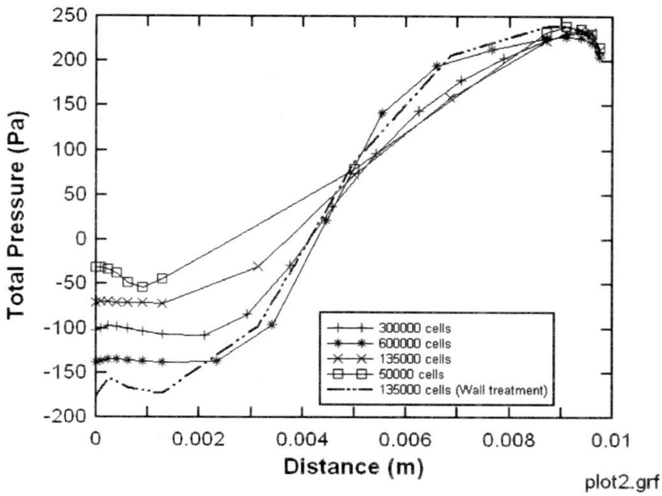

Figure 5: Influence of grid size of predicted results (Pressure profile at the entrance to third channel), standard k-ε model used in calculations.

4 Results and discussion

In order to assess the quality of the results, simulations were carried out using two higher order schemes: a second-order upwind method and the QUICK method. The resulting system of non-linear algebraic equations was under-relaxed and solved by the SIMPLE algorithm. The iteration procedures started with investigation of pressure drop (i.e., test without heating) until the convergence criteria were reached. Thereafter the energy equation was turned on and the analysis was further continued until convergence, Figure 4. The parameters such as the heat transfer coefficient, temperature at the outlet, mass flow rate, turbulence y^+ values were monitored during the simulations to better control the convergence of solutions. Each analysis consumed totally 4 hours (2000 iterations) on Intel Core 2 Duo 2.4 GHz computer with 2 GB RAM memory with memory speed of 667MHz dual channel.

Due to the lack of published experimental data in the literature, for simultaneously developing flow in rectangular ducts only fully developed friction factor and Nusselt number are used for comparison with the model.

The friction factor behaviour for the investigated cases of channel Reynolds number can be seen in Figure 7. The friction characteristic of the linear channels monolith seems to have a magnitude between the characteristics of the laminar [6], and checkerboard [3] cases. As expected, the friction factor is lower than for the laminar case with some turbulence intensity existing at the inlet boundary. The fact that the present design avoids a complex distributor plate causes the pressure losses to be lower than those for the case of the checkerboard monolith.

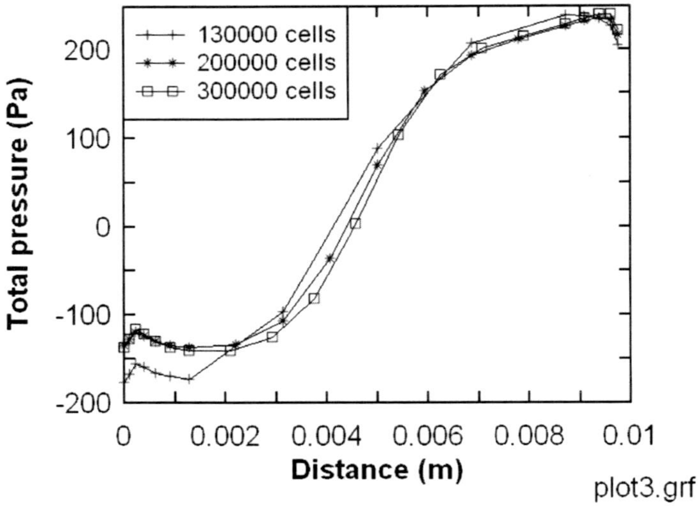

Figure 6: Influence of the grid size on the predicted results (Pressure profile at the entrance to third channel), enhanced wall treatment model used in all calculations.

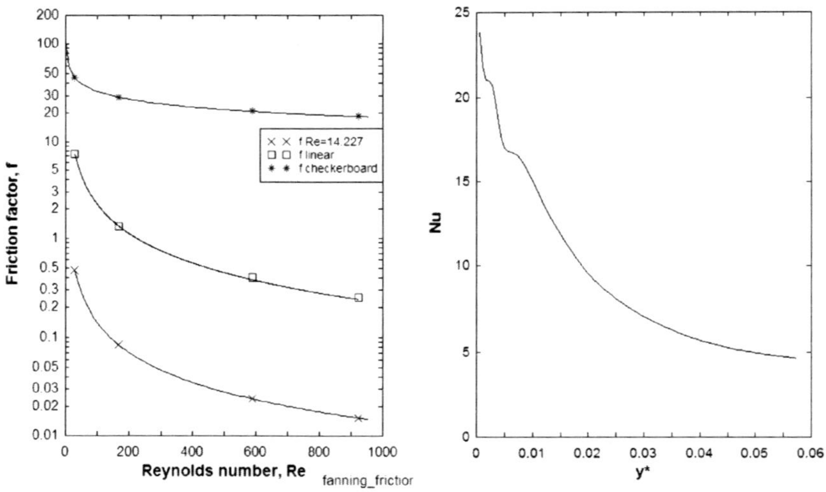

Figure 7: Left: Flow friction characteristics. Right: Nusselt number along the main flow direction, Re=160 (data taken from inlet to outlet of the first channel).

The correlation between the friction factor and Reynolds number has been found to:

$$f = 218.4\, Re^{-0.995} \tag{10}$$

It is valid for: $30 < Re < 1000$. Hydrodynamic and thermal boundary layer development caused by forced convection makes Nusselt number $\overline{Nu}$ to decay in the downstream increasing y^*, Figure 8. High values of Nusselt number at the channel entrance are caused by the driving potential for heat transfer of cold fluid at the entrance into the channels. At some position y* along the channel Nu approaches its fully developed value which is calculated to 4.7. As expected this value is higher than the laminar case because of flow regime is in the transition zone. The following formula is an empirical fit of data obtained from numerical simulations:

$$\log Nu = 1.366 - 21.86\, y^* + 121.78\, (y^*)^2 + 1506.8\, (y^*)^3 - 11224.7\, (y^*)^4 \qquad (11)$$

Typical longitudinal decrease of Nu number with increasing y* can be found, (see Figure 9). The thermal boundary layer development is disrupted by the two side walls which are not contributing to the heat transfer and also by the non-uniform fluid velocity (see Figure 11).

At the entrance of the each channel a vortex motion increases the heat transfer and fluid temperature as can be observed from Figure 12. This effect diminishes with increasing y*. The same effect explains the curve trends in Figure 10 and Figure 11.

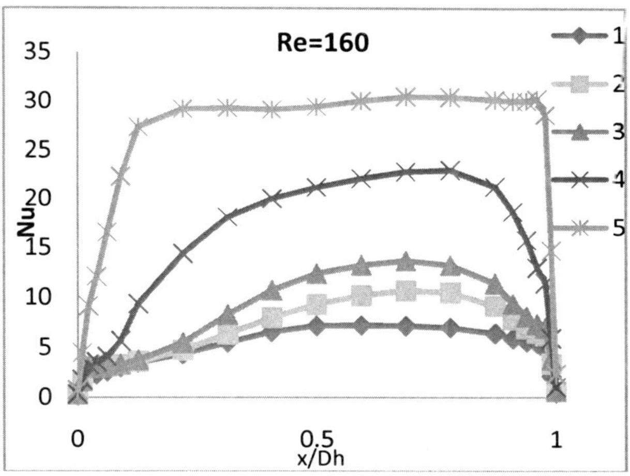

Figure 8: Spanwise distribution of Nusselt number Nu into Channel 1 (curve 1, y*=0.0619; curve 2, y*=0.040; curve 3, y*= 0.032; curve 4, y*=0.014; curve 5, y*=0.005.

5 Conclusions

Numerical simulations of developing flow in the transition region and heat transfer in a five monolithic square channel were presented for various Reynolds numbers. The current 3D CFD simulation was applied on vertical channels arranged by linear flow arrangement where constant wall temperature was

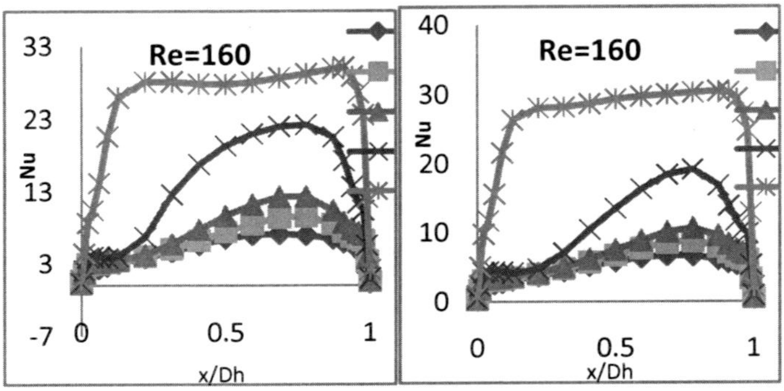

Figure 9: Left: Spanwise distribution of Nusselt number Nu in Channel 3 (curve 1, y*=0.0619; curve 2, y*=0.040; curve 3, y*= 0.032; curve 4, y*=0.014; curve 5, y*=0.005). Right: Spanwise distribution of Nusselt number Nu in Channel 5 (curve 1, y*=0.0619; curve 2, y*=0.040; curve 3, y*= 0.032; curve 4, y*=0.014; curve 5, y*=0.005).

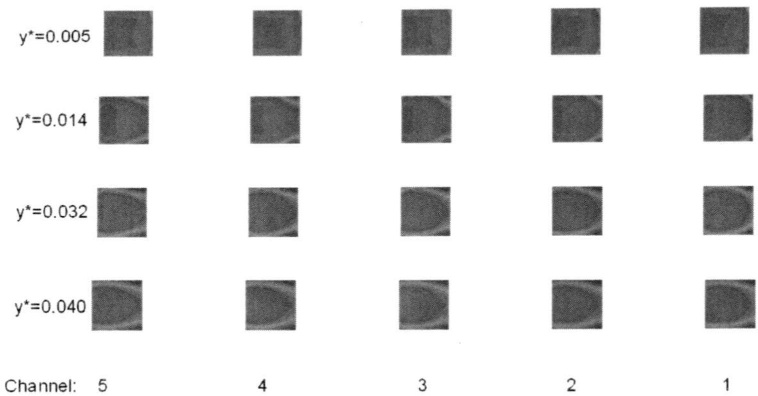

Figure 10: Axial temperature contour plot at different y* locations in all 5 channels.

applied on two walls and thermal insulation on the other two walls. From the structure of gas flow and temperature field it was found that the generated vortex at the entrance of each channel caused disruption of both the hydrodynamic and thermal boundary layer. Further, the total pressure drop was found to be somewhat higher than the values from the literature for laminar flow but lower for a monolithic exchanger where the channels have been arranged in a checkerboard flow arrangement. Also a correlation for the heat transfer coefficient was established and the values are higher than for the strict laminar flow. This was explained by the transition flow regime operation of the

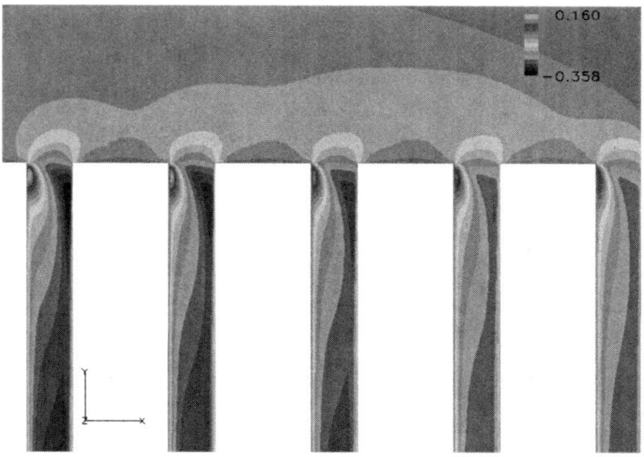

Figure 11: v-velocity contour plot located at the middle of z-plane (Observe negative prefix for downward velocity).

exchanger. Finally, this study might be regarded as an improved modelling procedure for gas flow and convective heat transfer in monolithic heat exchangers/reactors.

References

[1] Wunsch, R., M. Fichtner, O. Görke, Katja-Haas-Santo, and K. Schubert. *Process of Applying Al_2O_3 Coating in Microchannels of Completly Manifactured Microstructured Reactors.*, Chem. Eng.Tech. 25, s. 7, 2002.

[2] Schmitt C., et al *Ceramic Plate Heat Exchanger for Heterogeneous Gas Phase Reactions,* Chem Eng. Technol., 28 (3), s. 337, 2005.

[3] Selimovic, F., Sunden, B. and Bruun, T. *Computational Analysis of Gas Flow and Heat Transport Phenomenain in Monolithic Structures for High Temperature Processes.* San Francisco: ASME Proceedings HT2005-72183, 2005.

[4] Fluent 6.3 User's guide. 2006.

[5] Javeri, V., *Laminar Heat Transfer In a Rectangular Channel For The Temparature Boundary Condition Of The Third Kind.*, Int. J. Heat and Mass Transfer, 21, ss. 1029–1034, 1977.

[6] Shah, R.K. and London, A. L. *Laminar Flow Forced Convection in Ducts.* u.o. : Academic Press, ISBN 0-12-020051-1, 1978.

[7] Incropera, F P and DeWitt, D P. *Fundamentals of Heat and Mass Transfer.* New York : Wiley, 1996.

Scalar characteristics of a lean premixed turbulent V-shape flame (air-butane)

M. S. Boulahlib, S. Boukebbab, I. Amara & E. Ferkous
Department of Mechanical Engineering,
Constantine University, Algeria

Abstract

This paper proposes an experimental study of the scalar fields of a lean V shape premixed turbulent flame. The thermometry with the thermocouple numerically compensated was used for determining the thermal fields of a premixed butane-air V shape flame. The structure of both the inert and reactive jet was studied for a Reynolds number of 7000. Turbulence is generated by a perforated grid situated upstream. The main results of the thermal field in the V shape flame have showed more important results and a representative form of the inner structure of the flame was then obtained. The study is to determine how the turbulence and equivalence ratio can affect the temperature field, the flame height, the variable progress and therefore the combustion process.
Keywords: experimental, thermocouple, compensated temperature, premixed combustion, butane, turbulence grid.

1 Introduction

Knowledge of the flame structure can be improved by a better comprehension of the flow turbulence phenomena and combustion. So, it is necessary to understand the physical and chemical processes dominating within the combustion system. In most practical applications, premixed turbulent combustion takes place in the laminar flamelet regime. This combustion regime is characterized by the combustion mode in which all length scales of the turbulent structures in the reacting flow are larger than the flame thickness.

In the flamelet regime, where one assumes a chemical reaction time lower than the turbulence time scale, the effects of the interaction between the phenomena of turbulence and combustion are represented at the same time by the

WIT Transactions on Engineering Sciences, Vol 61, © 2008 WIT Press
www.witpress.com, ISSN 1743-3533 (on-line)
doi:10.2495/HT080041

local and global structure of the flame. The flame is folded by the turbulent flow, which leads to an increase in its surface. Simultaneously, the local structure of the flame is modified by the flame front orientation and shearing of the turbulent flow represented by the flow divergence in the tangent plan flame. Poinsot et al. [12], based on numerical results obtained from DNS, estimate in the flamelet regime that the small structures are able to penetrate and modify the internal structure of the flame front. Peters [11], supposes that the small structures can penetrate the preheating zone, but that they are unable to disturb the fine reaction zone inside the flame front. The preliminary steps of measuring the dynamic and scalar structure using a V shape flame have been performed by a number of researchers. Cheng et al. [4] investigated the conditional velocity statistics in premixed CH_4 -air and C_2H_4-air turbulent flames. Miles and Gouldin [7] developed and applied an experimental technique for determining the mean rate of chemical reaction <w>, in turbulent V-shape flames. The influence of free stream on the statistics of a turbulent premixed V-flame is numerically investigated by Chan et al. [3] and is compared to experimental results. Knaus et al. [5], in an article using a new technique for obtaining instantaneous, high resolution, three dimensional thermal structure data from turbulent flames, crossed-plane Rayleigh imaging. This technique is used to measure temperature gradient data for a lean, premixed methane-air turbulent V-flame. Recently Tang and Chan [15], using a contour advection numerical model, have investigated the effect of turbulence on various quantities such as flame brush thickness and flame surface density.

One of the objectives of this paper concerns the characterization of flame structure for a premix in terms of the flame height, temperature field and mean progress variable. The simultaneous effect of several scales of turbulent flow on the flame front was studied. Stationary flames were recorded and treated to describe the space properties for various turbulence conditions and equivalence ratios. This quantity is estimated in order to characterize the average interaction between the flame scalar field and dynamic one in which it is propagated.

2 Experimental device

The experimental device used consists of a turbulent burner where the premixing to butane/air is convected with a mean velocity of 6 m/s (Boulahlib et al. [1]). The burner includes a feeding system of butane, which is injected axially into a premixer with air. A laminarisation system, and a set of grids perfectly dimensioned according to the expected turbulence, allows the control of the turbulence level. Spaces between the holes give rise to wakes that mix downstream from the grid. In the absence of a turbulence grid, the outgoing flow of the burner is on the level of very weak turbulence (lower than 1%). The burner exit section has a length of 180mm and a width of 6mm.

The grid turbulence is well-known and it is the only one that is connected with a homogeneous and isotropic turbulence. Interchangeable grid turbulence can generate various conditions of homogeneous and isotropic turbulence: laminar case and turbulence intensity is 6%, 6.4% and 7.5%. Thus, the flames

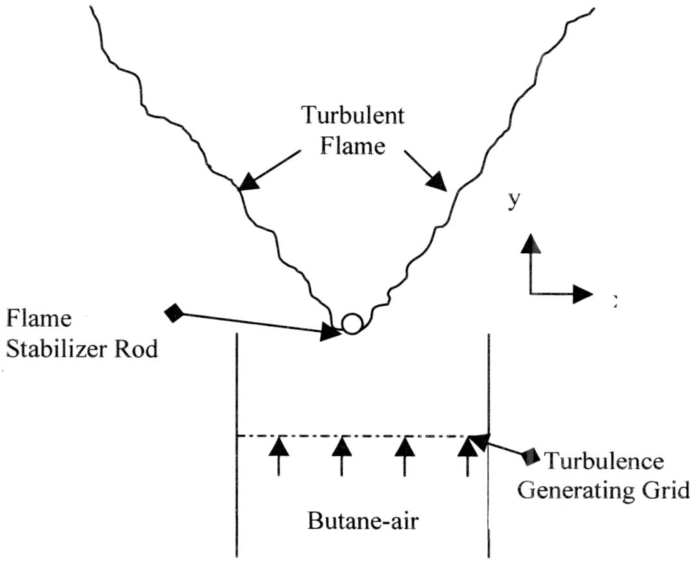

Figure 1: V–anchorage flame geometry.

produced with various conditions of equivalence ratio (equivalence ratio = 0.9 ÷ 1.4) and turbulence is propagated in the combustion chamber under atmospheric pressure (Figure 1). The characteristics of the turbulent flow like velocities were determined by Hot Wire. Table1 summarizes the turbulence characteristics and combustion parameters for the various flames displayed. The ratio u'/S_L, where S_L represents the speed of non-stretched laminar flame (figure 3), allows one to combine the turbulence and combustion conditions. The values of u'/S_L vary from 2.46 to 4.04. The studied flames are located in the flamelet regime.

Table 1: Turbulence conditions.

Grid	M(mm)	d(mm)	d/M	σ(%)	C_D (%)	U(m/s)	u'(m/s)	u'/U (%)
P	2.40	2	0.83	0.38	37.5	5.75	0.425	7.4
M	3.52	3	0.85	0.34	26.5	5.68	0.370	6.5
G	4.54	4	0.88	0.30	18.4	5.28	0.318	6.0

2.1 Temperature measurement by thermocouples in a reacting flow

A thermocouple is made of two wires of different types whose ends are connected by joints. When one of the junctions is maintained at a temperature different from the others, there appears in the electrical system an electromotive force due to the different nature of metals (the Seebeck effect). This electromotive force is a function of the temperature difference between the two junctions. In our study we use thermocouple type K. The hot weld of these thermocouples is made with two wires having small-diameters so as to increase the temporal and spatial resolutions of the measurement. The wires are fixed on pins that are more resistant to flow. Each pin is made of the same alloy as the corresponding part of the wire. The use of thermocouples for a temporal resolution of the measurement of high temperatures requires specific treatment of these values. When the temperature of the environment is high, the loss of radiant heat from the wires becomes important. Thus, the temperature measured by the hot junction is less than the temperature of the environment. These losses have been estimated and discussed by many authors (Scadron and Warshawsky [14], Bradly and Matthews [2], Paranthoën and Lecordier [10]). Their estimates by one of the proposed models can correct the difference between the thermocouple temperature and the actual temperature of the gas. In addition, the temporal resolution of the thermocouple signal is a very important parameter for studies of fluctuating flows. The thermocouple has been used in numerous applications of combustion. Neveu et al. [8] have coupled the measurement of temperature (2 kHz) and speed obtained simultaneously to correlate and examine the turbulent flow of a non-pre mixed methane/air burner. Poireault [13] studied a 'swirl' type burner, in order to understand the influence of the number of swirls on the stabilization of the flame. He characterized the areas of the three-dimensional flow through the modality of the probability function of the temperature signal. Larass [6] studied the thermal and dynamic fields in a domestic model boiler.

2.1.1 Heat exchange

The theoretical study of heat transfer between a thermocouple and the environment was carried out by Scadron and Warshawsky [14]. The study takes into account signal compensation for the losses by radiation and quantifies the response time of the thermocouple. There are also other studies that discuss the theoretical approach of this method and show its advantages and its limit (Nina and Pita [9], Neveu et al. [8]). The energy balance performed on a unit length dx of the thermocouple in a flow is presented below (eqn (1)). It is valid for an ideal welding of the fine wire of the thermocouple. This is considered as a cylindrical body.

Accumulation (Acc) = Convection (A) + Conduction (B) + radiation (C) + Catalysis (D)

$$\underbrace{\pi r^2 dx \rho c_p \frac{\partial T_c}{\partial t}}_{Acc} = \underbrace{2r\pi dx h\left(T_g - T_c\right)}_{A} - \underbrace{r^2 \pi dx \lambda_f \frac{\partial^2 T_c}{\partial x^2}}_{B} + \underbrace{2r\pi dx \varepsilon_c \sigma\left[\varepsilon_{pa} T_{pa}^4 + \varepsilon_g T_g^4 - T_c^4\right]}_{C} + \underbrace{2r\pi dx Q_{cat}}_{D} \tag{1}$$

2.1.2 Digital compensation

We recall that compensation is needed for two reasons. The first relates to the temperature displayed by a thermocouple, which is not necessarily the gas temperature of the environment, as a result of losses by radiation from high temperature thermocouples. For the gas temperature, it is necessary to estimate and correct those losses. The second reason relates to the response time of the thermocouple, which is limited by the inertia of two wires.

$$\frac{\partial T_c}{\partial t} = \frac{4Nu\lambda_g}{d^2 \rho c_p}\left(T_g(t) - T_c(t)\right) \tag{2}$$

$$\text{Or: } \tau_{cv} = \frac{d^2 \rho c_p}{4Nu\lambda_g} \tag{3}$$

$$T_g = T_c + \tau_{cv}\left(\frac{\partial T_c}{\partial t} - \frac{\varepsilon_c \varepsilon_{pa}}{1-(1-\varepsilon_c)\cdot(1-\varepsilon_{pa})} \cdot \frac{4\sigma\left(T_c^4 - T_{pa}^4\right)}{d\rho\rho_p}\right) \tag{4}$$

The equation obtained allows us to have simultaneously offsetting losses by radiation and inertia (eqn (4)). It is mainly a function of temperature measured by the wire and the Nusselt number. It should be noted that the Nusselt number is a function of the Reynolds number and therefore depends on the temperature of the gas.

3 Results

3.1 Flame heights

One of the items considered desirable in a data base on turbulent flames is the measurement of some macroscopic variables such as flame height (or length). They constitute important major elements to define the global spatial properties of the turbulent flames. The height of the flame H is defined as the value where the value of temperature is at maximum. Analyses of images of spontaneous emission of flame showed a reduction of the combustion process with the increase in the turbulence intensity. Indeed it is accompanied by a decrease from the height of the flame and can affect the average field and unruly directly on the flame height and the combustion process (Figure 2).

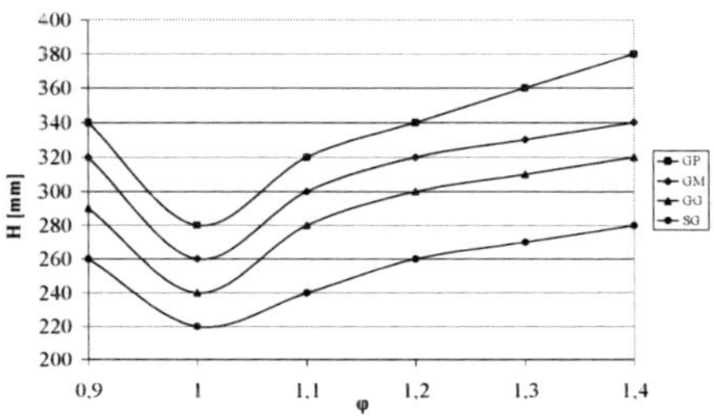

Figure 2: Flame heights with the equivalence ratio and turbulence.

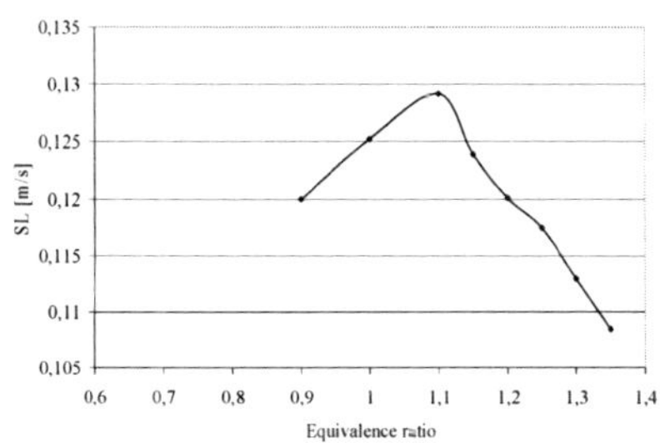

Figure 3: Laminar flame velocity.

Regarding equivalence ratio, the increase from the lean burn, we can see that for various equivalence ratio worth of the flame height, all profiles vary in terms of sizes around the stoichiometry. In fact, S_L decreases, with the equivalence ratio for Ø<1 (Fig3), and height flame increase significantly for the same mass flow. The flame heights allow us to highlight the existence of three areas. A first region of lean burn is characterized by large flame heights and low wrinkling, a second area corresponds to the location of the stoichiometric line with a significant reduction in the flame height and an increase in the wrinkling, and a final area of high equivalence ratio where there is a noteworthy increase in the size of the flames and decrease wrinkling. There is a discrepancy of these minimum sizes of the change in equivalence ratio inward depending on the increase in the turbulence intensity; this is explained by the interaction between combustion and turbulence.

3.2 Laminar temperature flame

Temperature profiles for laminar flames measured are shown in figures 4 and 5, where temperature is plotted against a spatial coordinates X. These data are measured using thermocouple and represent the average of 10 temperature fields. The evolution of the mean temperature profiles with different equivalence ratios and inlet velocities are carried out.

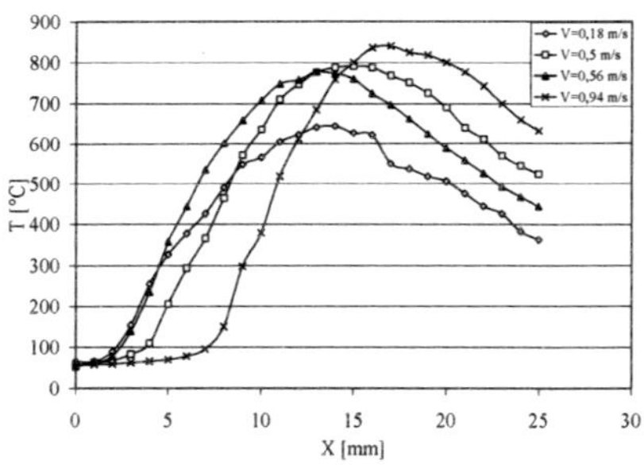

Figure 4: Effect of inlet velocity on temperature field.

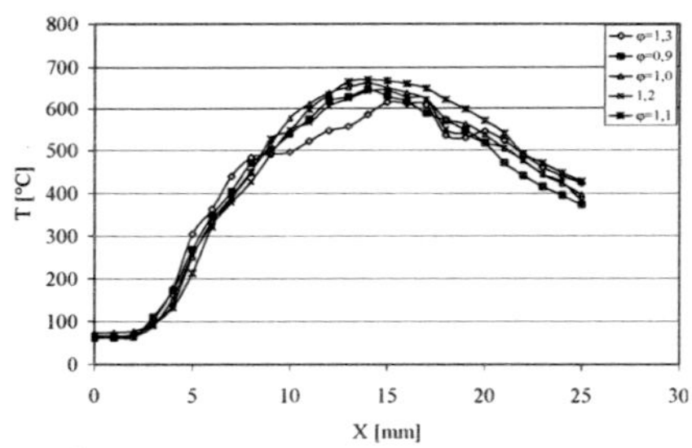

Figure 5: Temperature with equivalence ratio.

3.3 Turbulent flame

Figures 6–8 thus show the profiles of temperature in the laboratory coordinates as function of y (various distances), φ (equivalence ratio) and different grid (turbulence intensities). The trends of the variations of these quantities are as expected. The original temperature data are first fitted with a cubic spline, and then smoothly interpolated to yield equally spaced data. The experimental data on the temperature show a significant expansion of the gas away from the centerline in the radial direction because heat release. Temperature profiles measured in the same burner at different distances should be similar, and weakly affected by turbulence characteristics and mixture composition. The areas of maximum temperatures are situated on the

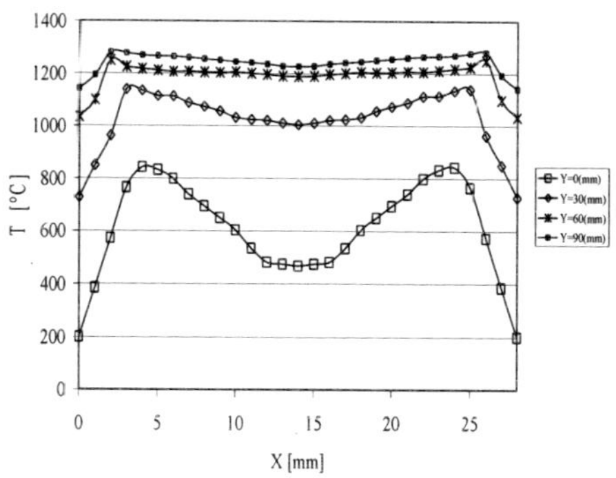

Figure 6: Flame temperature as a function of y for φ = 0.9 and P grid.

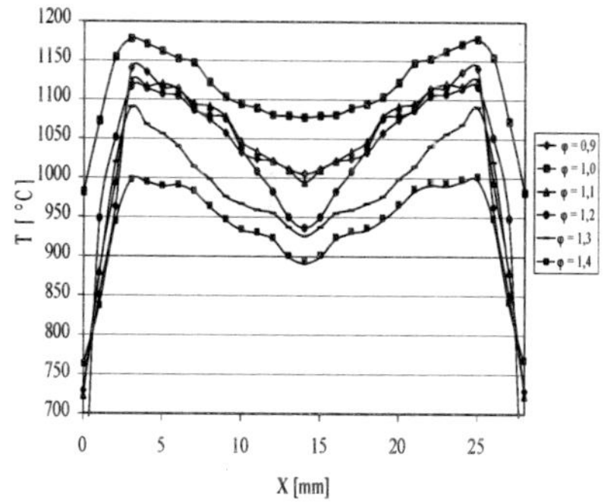

Figure 7: Mean temperature profile equivalence ratio at the burner exit.

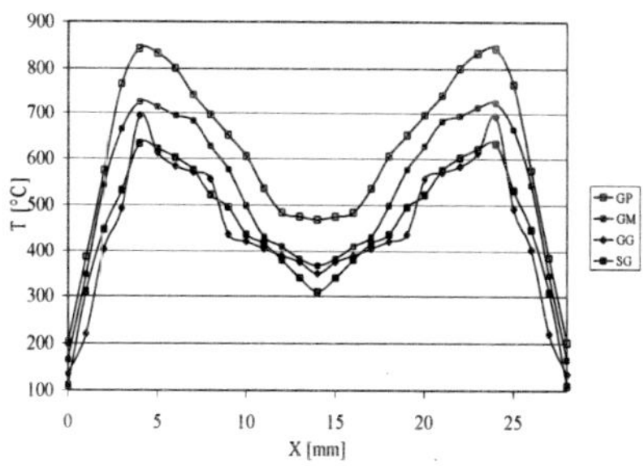

Figure 8: Temperature for various turbulence intensities at the burner exit plane.

banks of the flame (burned gas), and are at their lowest in the center (fresh gas). The fresh gas is kept at temperatures between 350 and 600°C for rich mixtures. This amount of heat gained by fresh gas is due to diffusion phenomena of free radicals and rich flames radiation. The minimum temperature at the outlet of the burner, then to grow quite quickly when one moves along the flame. The temperature is constant in the flame front from a certain height. Beyond this height the temperature remains relatively constant across the flame (only burn gas), on account of the turbulent kinetic energy that reaches its maximum value, just outside the jet, this is due mainly to the presence of a thermal gradient between the velocity jet of burned gas, and ambient air, which represents in terms of production the turbulent kinetic energy, the latter will suffer a decline throughout the flame, the influence of combustion is clear. We can note figure 6 that gradient are very close to those of upstream boundary of the luminous zone, with the spreading representing the uncertainties caused by flame flickering. The only slight deviation from the above observation is for the near y = 90 mm, for which the final, peak temperature at the stagnation flame is upper than other case. Figure 8 shows the temperature profiles in the laboratory coordinates for the different equivalence ratio. Also indicated for reference are two important peak locations, which respectively correspond to the maximum temperature gradient, maximum heat release. It is of interest to note that while the deep violet radiation associated is with the luminosity, if lean hydrocarbon-air-mixtures it is due to excited CH radicals. By varying equivalence ratio (φ = 0.9 to 1.4) and for each grid, we find that the maximum temperatures in the flame can reach the adiabatic temperature in the area stoichiometric (φ = 1) and the highest intensity of turbulence (P grid), then it decreases by increasing equivalence ratio and reducing turbulence. Figures 9 further show that whiles the flame moves closer to the surface when subjected to increasing turbulence intensities. For further scrutinize the possible sensitivity of the flame structure to turbulence intensities, the profile of temperature in the flame coordinates are compared. Recognizing that profile comparison provides an even more stringent test of sensitivity than profile comparison, the results clearly show the existence of such sensitivity. For the same station and the same equivalence ratio, we see that turbulence intensity increase the temperature.

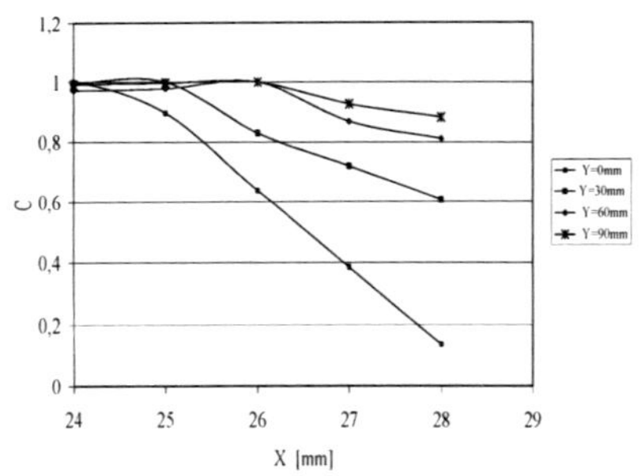

Figure 9: The field of mean variable in the flame brush.

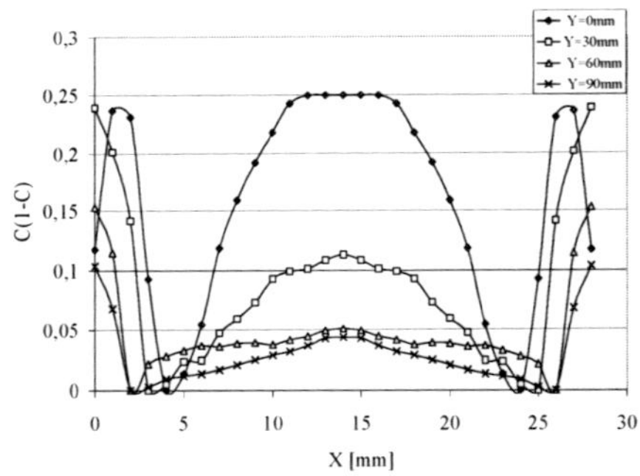

Figure 10: Fields of progress variable variance $C'^2 = C(1-C)$.

The mean progress variable is defined as follow: $C = (T-T_u)/(T_b-T_u)$, where T_b is the adiabatic flame temperature, and $T_u = 298K$. The evolution of the progress variable along the axis of V-shape flame is shown in figure 9. The results of variable progress <C> can be extracted from the temperature fields. The increase of the mean progress variable, <C> along the axial direction is consistent with the flame temperature, and they increase monotonically with X. The scalar profiles of figure 9 are typical of wrinkled flames found in other flame configuration such as stagnation flow stabilized flames, V flames and large conical flames (Cheng et al. [4]). The main differences are the flame zone thickness. The profiles are not perfectly symmetrical this is probably an imperfection of the burner. The curves of $<C'^2>$ (figure10) can then be deduced from the ones of <C>. These results are interest for comparisons with prediction methods. They will also be directly useful for further discussions though comparisons with other quantities.

4 Conclusion

From the temperature fields, the flame heights were determined. The latter allows one to show the existence of three zones: an area of lean flame characterized by important heights of flame and a weak wrinkling. A second zone corresponds to the localization of the stoichiometric line with a significant reduction in the height of the flame and an increase in wrinkling. The last zone of high equivalence ratio is where an increase in the size of the flames and a reduction in wrinkling are noticed. One observes a shift of these minimal sizes towards the interior as a function of the turbulence intensity increase. The overall reaction is exothermic combustion, acceleration are usually corresponds to an increase in temperature in the reaction zone. More precisely this reaction induced considerable heat release in a very thin zone leading to very high temperature gradients. The equivalence ratio plays a very important role in terms of the quantity of heat release, its variation affects different parameters. Turbulence raises the temperature, so the turbulence improves combustion. The compensation digital improves the measurement results.

References

[1] Boulahlib M.S., Renou B., Boukhalfa A. and Nemouchi, Z., "Caractérisation expérimentale de la structure dynamique et scalaire d'une flamme turbulente de prémélange CH_4-air: par PIV et tomographie Laser, Sciences & Technologie N°22B, pp 67 78, décembre 2004.

[2] Bradly D. and Matthews K. J., "Measurement of high gas temperatures with wire thermocouple", Mechanical Engineering Science, Vol. 10 (4), 1968.

[3] Chan C.K., Wang H.Y. and Tang H.Y "Effect of intense turbulence on turbulent premixed V-flame" International Journal of Engineering Science, 41 pp. 903 916, 2003.

[4] Cheng R.K., L. Talbot and Robben F., "Conditional velocity statistics in premixed CH_4-air and C2H4-air Turbulent Flames" 20th Symposium (international) on Combustion, The Combustion Institute, pp. 453 461, 1984.

[5] Knaus D.A., Sattler S.S., and Gouldin F.C., "Three-dimensional temperature gradients in premixed turbulent flamelets via crossed-planed Rayleigh imaging" Combustion and Flame" 141: pp. 253 270, 2005.

[6] Larass N., "Caractérisation expérimentale des champs thermiques et dynamiques de la combustion dans une chaudière domestique modèle", Thèse de Doctorat, Université de Rouen, 2000.

[7] Miles P.C. and Gouldin F.C., "Mean reaction rates and flamelet statistics for reaction rate modeling in premixed turbulent flames" 24th Symposium

(international) on Combustion, The Combustion Institute, pp. 477 484, 1992.

[8] Neveu F., Corbin F., Perrin M. and Trinité M., "Simultaneous velocity and temperature measurements in turbulent flames", 7th International Symposium on Applications of Laser Techniques to Fluid Mechanics, Lisbon Portugal, 1994.

[9] Nina M. and Pita G., "Measurements of fluctuating gas temperatures thermocouples", AGRAD conference proceedings, panel symposium, Bergen, Norway, pp. 32.1 10, 1985.

[10] Paranthoen P., Lecordier J.C. "Mesures de température dans les écouements turbulents", Revue Générale de Thermique pp. N°413, pp. 283 308, 1996.

[11] Peters N., « Laminar flamelet concepts in turbulent combustion » 7th Symposium (International) on Combustion, The Combustion Institute, pp. 461 468, 1986.

[12] Poinsot T., Veynante D., & Candel S. "Quenching processes and premixed turbulent combustion diagrams", J. Fluid Mech., vol: 228, pp.561–606, 1991

[13] Poireaut B., "Mécanisme de combustion dans un brûleur méthane air de type 'swirl' Influence de l'intensité de la rotation", Thèse de Doctorat, Université de Poitiers France, 1997.

[14] Scadron M. D. and Warshawsky I, "Experimental determination of the time constant and Nussell numbers for bare-wire thermocouples ", National Advisory Committee for Aeronautics, Technical note 2599 Washington, 1952.

[15] Tang B.H.Y., and Chan C.K., "Simulation of flame surface density and burning rate of a premixed turbulent flame using contour advection" Combustion and Flame" 147, pp. 49 66, 2006.

Numerical simulation of incompressible laminar fluid flow in tubes with wire coil inserts

D. Muñoz-Esparza[1], J. Pérez-García[1], E. Sanmiguel-Rojas[2], A. García-Pinar[1] & J. P. Solano-Fernández[1]
[1]*Department of Thermal and Fluids Engineering, Technical University of Cartagena, Spain*
[2]*Department of Mechanical and Mining Engineering, University of Jaén, Spain*

Abstract

Numerical simulations of steady incompressible laminar 3D fluid flow with heat transfer in smooth round tubes with wire coil inserts have been accomplished. Four wire coils with different geometrical characteristics of dimensionless pitch *(p/d)* and wire-diameter *(e/d)* have been studied for Reynolds numbers lower than 500–700 using the commercial package software FLUENT.

The numerical results have been compared with experimental data previously obtained by the authors and a good agreement is reached. These results allow one to know the internal flow structure, to identify the critical Reynolds number of oscillating laminar flow and to obtain practical correlations of the Fanning friction factor. So, the internal flow structure under isothermal conditions has been analyzed and a set of practical correlations of Fanning friction factor have been obtained. The heat transfer enhancement produced by the insertion of helical-wire-coils has also been treated. The preliminary results obtained for uniform heat flux are showed. The developed numerical methodology can be used to assess the heat transfer enhancement produced by other kinds of insert devices, to reduce the experimental efforts and to obtain more reliable correlations for heat exchangers design process.

Keywords: wire coil inserts, pressure drop, heat transfer enhancement, heat exchangers.

WIT Transactions on Engineering Sciences, Vol 61, © 2008 WIT Press
www.witpress.com, ISSN 1743-3533 (on-line)
doi:10.2495/HT080051

1 Introduction

Heat transfer processes of viscous fluids usually take place in laminar or transitional regimes, where transfer rates are particularly low. Heat exchangers working under these flow conditions are usually candidates to undergo an enhancement technique. Among the different techniques that are effective to improve the thermo-hydraulic behaviour in the tube-side in single-phase laminar flow, the insert devices stand out. The dominant literature (Bergles [1], Webb and Kim [2]) usually mentions five types: wire coils, twisted tapes, extended surface devices, mesh inserts and displaced elements. One of the main advantages of insert devices with respect to other enhancement techniques such as the artificial roughness by mechanical deformation or internal fin types is that they allow an easy installation in an existing smooth-tube heat exchanger.

Due mainly to their low cost, the insert devices which are most frequently used in engineering applications are wire coils and twisted tapes. Wire coil inserts are devices whose reliability and durability are widely contrasted. This is a cheap enhancement technique which is completely viable for many industrial applications, however, the studies in laminar regime are scare in open literature. This fact hinders a widespread use of wire coil inserts in industrial heat exchangers.

It is essential to understand the flow pattern in order to determine how the different flow structures induced by the wire coil augment the pressure drop, because the global improvement is based on the *heat transfer/pressure drop* ratio. A first experimental approach was carried out by Ravigururajan and Bergles [3] using injection of colored dye. Garcia *et al* [4] made a more complete study using hydrogen bubbles visualization technique and they stabilised the flow pattern qualitatively.

The necessity to obtain more accurate results and a wider range of information than qualitative visualization techniques has led in an effort to develop a numerical methodology that allow to study the flow pattern thoroughly. Mokamati *et al* [5] tried to carry out this task in turbulent flow but they were not successful.

In this work, a numerical methodology to study the thermo-hydraulic characteristics in laminar regime has been developed. Numerical results have been compared with experimental data that have allowed to validate the numerical methodology. Friction factor correlations have been obtained and have been compared with experimental data previously obtained by the authors for laminar region [6]. Flow pattern has been analyzed and compared with visualization results, and finally, preliminary results of heat transfer are being obtained.

2 Numerical simulation methodology

2.1 Mathematical model

The fluid flow studied is governed by 3D incompressible steady-state form of the Navier-Stokes equations. They may be written in Cartesian tensor notation as:

$$\frac{\partial}{\partial x_i}(\rho u_i) = 0 \tag{1}$$

$$\frac{\partial}{\partial x_i}(\rho u_i u_j) = -\frac{\partial p}{\partial x_i} + \frac{\partial}{\partial x_j}\left[\mu\left(\frac{\partial u_i}{\partial x_j} + \frac{\partial u_j}{\partial x_i}\right) - \frac{2}{3}\mu\delta_{ij}\frac{\partial u_k}{\partial x_k}\right] = -\frac{\partial p}{\partial x_i} + \frac{\partial \tau_{ij}}{\partial x_j} \tag{2}$$

$$\frac{\partial}{\partial x_i}(\rho u_i E) = -\frac{\partial (p u_i)}{\partial x_i} + \frac{\partial (\tau_{ij} u_i)}{\partial x_j} + \frac{\partial}{\partial x_i}\left(k\frac{\partial T}{\partial x_i}\right) \tag{3}$$

These equations are solved using FLUENT software, applying the finite volume method to the computational domain described in the following section.

2.2 Geometry and mesh

The computational domain is composed by a round smooth tube and a wire coil inserted. The resultant arrangement (a) and a sketch of the wire coil parameters in a longitudinal section (b) are shown in fig. 1.

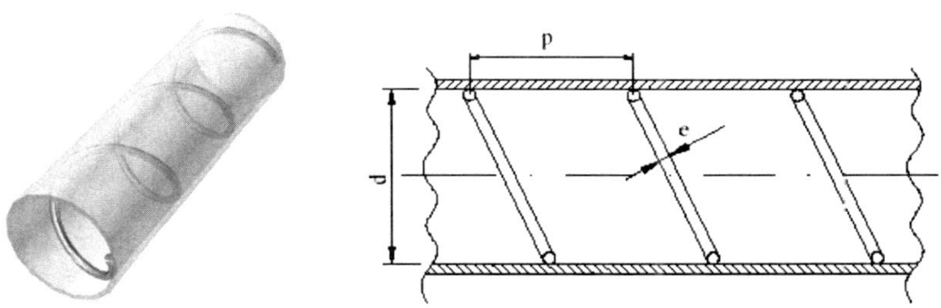

(a) (b)

Figure 1: (a) Studied geometry. (b) Sketch of a wire coil inserted inside a smooth round tube.

The numerical study was carried out on four wire coils fitted in a smooth tube. Table 1 shows geometrical parameters of the wire coils analyzed.

A structured, non-uniform grid has been built through meshing software Gambit. The most important gradients are expected to appear in surroundings of the wire coil surface. For this reason, the region close to the wire has been meshed with a high density of elements and a coarse mesh has been set out for the rest of the section. Simulations have been accomplished to 16 pitches tube lengths meshes for all the wire coils analyzed.

Table 1: Geometrical characteristics of the wire coils studied.

Wire coil	*d (mm)*	*p/d*	*e/d*	*p/e*
M01	18	1.17	0.074	15.76
M02	18	2.68	0.081	33.32
M03	18	1.86	0.099	18.75
M04	18	1.41	0.101	13.91

In order to ensure accurate results, a grid dependence study has been performed using the Richardson extrapolation method improved with Roache's Grid Convergence Index [7]. This study has allowed to determine the uncertainty of numerical results and the optimum mesh taking into account computational cost and reliability of results simultaneously. The uncertainty propagation has been evaluated according to the procedure detailed in [8]. Expanded uncertainty calculations based on a 95% confidence level showed limit values of 2.65% for Reynolds number, 1.44% for friction factor and 3.32% for Nusselt number.

2.3 Simulation hypothesis

The governing equations have been solved using FLUENT software. The segregated solver and implicit formulation have been chosen to study the flow behaviour under isothermal conditions. SIMPLE, QUICK and first-order upwind schemes have been used in order to discretize pressure-velocity coupling, momentum and energy equations. Non-isothermal flow conditions have been solved using coupled solver and implicit formulation with a second-order flow discretization.

Tube wall inner surface and wire coil were ascribed no-slip and adiabatic boundary conditions for the isothermal flow type. For non-isothermal flow condition, constant heat flux of $q''=11300\ W/m^2$ over the tube wall inner surface was applied. A free stream flow with uniform temperature and velocity profiles was imposed on the inlet transverse plane and zero diffusion flux was applied to the exit transverse plane for both cases. Propylene-glycol has been used in the numerical simulations, being its physical properties modelled as temperature functions.

3 Results and discussion

3.1 Friction factor results

Fanning friction factor has been obtained by means of the following expression:

$$f = \frac{\Delta p d^5 \pi^2 \rho}{32 L \dot{m}^2} = \frac{\frac{\Delta p}{L} d^5 \pi^2 \rho}{32\left(\rho U \frac{\pi d^2}{4}\right)^2} = \frac{\frac{\Delta p}{L} d^3 \rho}{2\mu^2\, Re^2} \tag{4}$$

Where $\Delta p/L$ is the pressure drop per unit length for fully developed flow, ρ is the fluid density and μ is the fluid dynamic viscosity. These values have been calculated using an area-weighted average in the transverse planes where the pressure drop has been evaluated.

Reynolds numbers from $Re=1$ up to transition Reynolds number to oscillating laminar flow, Re_{trans}, have been analyzed. Friction factor shows a linear tendency in double-logarithm axis. For this reason, it is interesting to obtain correlations for friction factor as a function of Reynolds number using a linear regression ($r^2=0.99$ for wire coil M03). Reynolds numbers lower than $Re=50$ are very improbable to appear in real situations and are not useful, therefore, numerical correlations have been obtained for *Re=50* to Re_{trans}. In fig. 2 numerical results, numerical correlation and experimental data are compared for wire coil M03. Differences lower than 2.5% have been obtained comparing numerical and experimental results for the same Reynolds number.

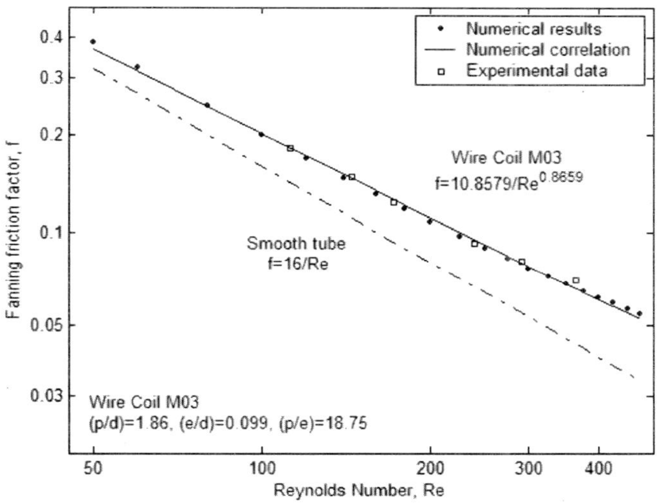

Figure 2: Numerical results, numerical correlation and experimental data for friction factor in wire coil M03.

Numerical correlations verify the form $f=a/Re^b$, where a and b are coefficients that depend on the wire geometry. The uncertainty level associated with the friction factor correlations, f, and the coefficients, a and b, have been computed according to [9]. Expanded uncertainty of 2.9% has been obtained for wire coil M03 friction factor correlation. Friction factor correlations for the four wire coils are summarized in table 2, with their uncertainties and range of application.

Table 2: Friction factor correlations in laminar regime for the wire coils studied.

Friction factor correlation	*Range of application*	*Uncertainty of correlation*
$f_{M01} = 9.2483/\mathrm{Re}^{0.8284}$	$50 < \mathrm{Re} < 500$	$u_{f,M01} = 3.5\%$
$f_{M02} = 11.6590/\mathrm{Re}^{0.9118}$	$50 < \mathrm{Re} < 700$	$u_{f,M02} = 2.6\%$
$f_{M03} = 10.8579/\mathrm{Re}^{0.8659}$	$50 < \mathrm{Re} < 475$	$u_{f,M03} = 2.9\%$
$f_{M04} = 9.5783/\mathrm{Re}^{0.8167}$	$50 < \mathrm{Re} < 475$	$u_{f,M04} = 3.6\%$

Heat exchangers do not work under isothermal conditions and it can be possible that this fact modifies the friction factor value. Numerical simulations applying a constant heat flux over the tube wall have been accomplished to assess this effect and friction factor has been calculated in this case, f^*. For low Reynolds numbers there are no differences between correlations, but when the Reynolds number is increased, the differences grow reaching a maximum value of 2.8% for Re_{trans}.

These results state that the correlations presented in table 2 are reliable for low Reynolds numbers and can be used for higher Reynolds numbers with a negligible error, which is of the order of its uncertainty.

3.2 Flow structure

The internal flow structure can be analyzed evaluating the velocity components in certain planes. The flow in a smooth round tube only has an axial component (theoretical Hagen-Poiseuille velocity profile), but the wire coil inserted modifies the internal flow structure. The wire coil induces a swirl effect on the main axial flow. In fig. 3, azimuthal velocity contours and vectors on a transverse plane are shown. It can be observed that the swirl effect is produced in the wire direction development (negative azimuthal velocities). In the region close to the wire coil surface the flow presents a small zone with an opposite swirl direction. This is due to the fact that main axial flow tries to return the flow that collides with the wire geometry to its previous position. The first swirl effect (negative azimuthal velocities) is more important and it affects to a cylindrical layer that grows towards the tube centreline when the Reynolds number increases. Wire geometries with small values of *(p/d)* imply a high swirl.

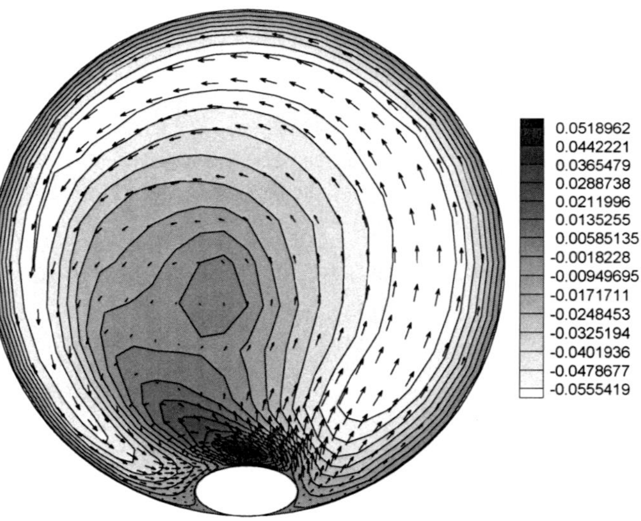

Figure 3: Azimuthal velocity contours (m/s) and azimuthal velocity vectors on a transverse plane for a Re=300 (flow entering on the sheet plane).

These results have been compared with experimental data obtained by Ravigururajan and Bergles [3] and García *et al* [4], concluding that numerical methodology reproduce the swirl effect appropriately and allow to obtain a wider range of information than with experimental techniques.

A recirculation zone just downstream of the wire coil appears for Reynolds numbers slightly higher than $Re=50$ and its characteristic size enlarges when the Reynolds number increases. The recirculation zone size depends principally on the wire-diameter, so, wire coils with high values of (e/d) will present larger recirculation zones. When the transition Reynolds number, Re_{trans}, is exceeded, the recirculation zone which has reached its maximum size, becomes unstable and the flow makes itself oscillating. This unstability is detected by means of numerical divergences that appear during the iterative process, as the algorithm used for the laminar zone is steady and this other region has a non-stationary behaviour. It has allowed to know the Re_{trans} for the set of wire coils studied, which depends on the wire geometry. For the smooth tube, this transition has been studied experimentally by García *et al* [4] and it is produced for $Re=1800$ approximately. The wire coil inserted advances the oscillating flow region to lower Reynolds numbers in the order of $Re=500\text{-}700$ for the wire coil geometries analyzed on this work. In fig. 4 axial velocity contours and vectors are shown for two different Reynolds numbers on a longitudinal plane. In fig. 4 (a) it can be observed that there are no recirculation zone for low Reynolds numbers ($Re<50$) and fig. 4 (b), where the recirculation zone takes place ($Re=300$). This structure agrees with experimental results showed in [4].

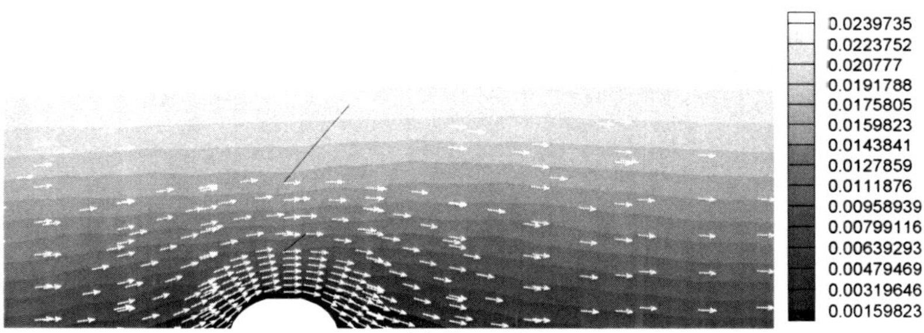

(a) Re=10

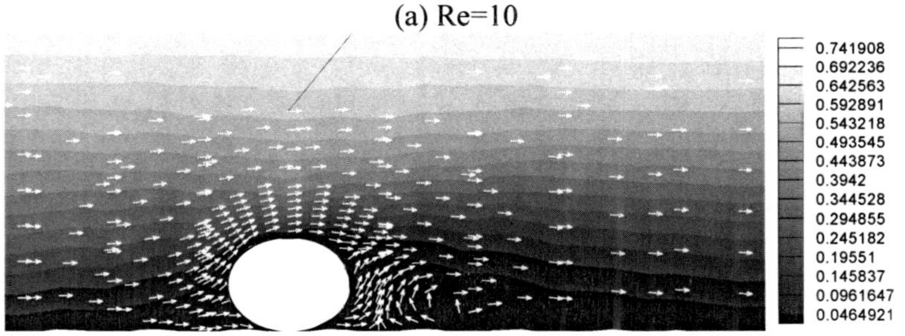

(b) Re=300

Figure 4: Axial velocity contours (m/s) and axial velocity vectors on a longitudinal plane. Detail round the wire coil geometry.

3.3 Heat transfer results

Preliminary results of heat transfer have been obtained. Nusselt number has been calculated from eqn. (5).

$$Nu = \frac{d}{k} \frac{q''}{T_{wall} - T_{flow}} \tag{5}$$

Where k, T_{flow} and T_{wall} are the mean values of thermal conductivity, flow temperature and inner wall temperature in a transverse plane, respectively. In fig. 5 numerical results of Nusselt number, experimental data and *Churchill* and *Ozoe* smooth tube correlation [10] are compared. Nusselt number is increased in the order of 1.7 times in respect with the smooth tube and this augment is almost independent of the Reynolds number for the laminar forced convection zone, showing these values a similar trend to the smooth tube. There are slight differences between numerical and experimental results, which are due to the different methodologies used to evaluate the Nusselt number.

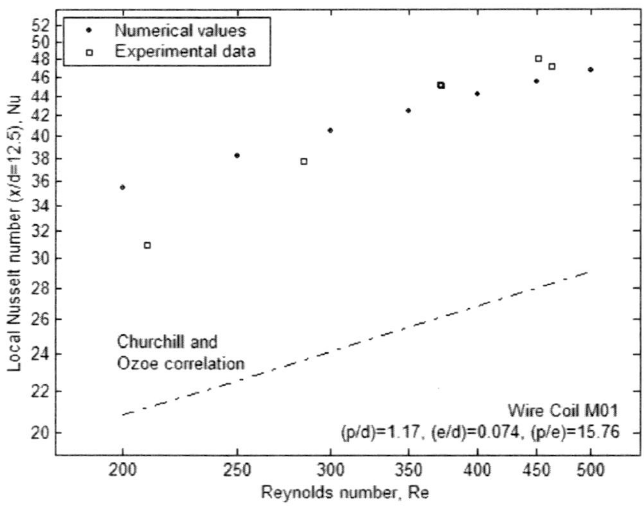

Figure 5: Numerical results and experimental data for the Nusselt number in wire coil M01 for a longitudinal dimensionless distance of *x/d=12.5*.

4 Conclusions

1. A numerical methodology to study the laminar fluid flow in tubes with wire coil inserts has been developed. The numerical results have been compared with experimental data obtained by the authors and a good agreement has been reached. Differences lower than 2.5% for the friction factor have been obtained comparing numerical and experimental results, which has allowed one to validate the numerical methodology.

2. The friction factor increases in reference to the smooth tube due to a coupled effect between the pitch *(p/d)* and the wire-diameter *(e/d)*. Small pitches and big wire-diameter geometries provokes an augmentation of friction factor for constant Reynolds numbers.
3. A set of practical and reliable correlations for the friction factor has been obtained as a function of the Reynolds number and their associated uncertainty has been calculated.
4. Simulations under isothermal and non-isothermal conditions have been made to study the effect of an applied heat flux over friction factor results. Slight differences appear between correlations, which are similar to the correlations uncertainties. By this reason, correlations showed in table 2 can be used in order to predict the friction factor with a negligible error.
5. The internal flow structure has been assessed. It is composed by a main axial flow with a swirl effect induced by the wire coil and a recirculation zone just downstream of the wire. The pitch determines the swirl effect, higher when it is reduced, and the wire-diameter determines the size of the recirculation zone, bigger when it is increased. The size of the recirculation zone enlarges and azimuthal velocity component rises when the Reynolds number augments. The recirculation zone becomes unstable for Reynolds numbers between 475–700 (Re_{trans}), depending on the wire coil geometry. These instabilities advance the transition region to lower Reynolds numbers in reference to the smooth tube.
6. Heat transfer has also been treated. Preliminary results show that wire coils generate a heat transfer enhancement but it is necessary to evaluate the global improvement taking into account the pressure drop. Different geometries are being studied and a set of correlations for Nusselt number will be obtained (in progress).

Acknowledgements

This research has been partially financed by the DPI 2007-66551-C02-01 grant of the ''Dirección General de Investigación del Ministerio de Educación y Ciencia de España''.

References

[1] Bergles, A.E., *Techniques to augment the heat transfer*, Handbook of Heat Transfer Applications, McGraw Hill, New York, 1985.

[2] Webb, R.L. & Kim, N.H., *Principles of Enhanced Heat Transfer*, second ed., Taylor & Francis Group, New York, 2005.

[3] Ravigururajan, T.S. & Bergles, A.E., Visualization of Flow Phenomena Near Enhanced Surfaces. *Journal of Heat Transfer (Transactions of the ASME)*, **116**, pp. 54–57, 1994.

[4] García, A., Solano, J.P., Vicente, P.G. & Viedma, A., Flow pattern assessment in tubes with wire coil inserts in laminar and transition regimes. *International Journal of Heat and Fluid Flow*, **28**, pp. 516–525, 2007.

[5] Mokamati, S.V. & Prasad, R.C., Numerical Simulation of Fluid Flow and Heat Transfer in a Concentric Tube Heat Exchanger. *Proc. of the 19th Canadian Congress of Applied Mechanics*: Calgary, 2003.
[6] García, A., Solano, J.P., Vicente, P.G. & Viedma, A., Enhanced of laminar and transitional flow heat transfer in tubes by means of wire coil inserts. *International Journal of Heat and Mass Transfer*, **50**, pp. 3176–3189, 2007.
[7] Freitas, C.J., Ghia, U., Celik, I., Roache, P. & Raad, P., ASME's quest to quantify numerical uncertainty. *41st AIAA Aerospace Sciences Meeting and Exhibit*: Reno, 2003.
[8] ISO, Guide to the Expression of Uncertainty in Measurement, first ed., International Organization for Standarization, Switzerland, 1995.
[9] Coleman, H.W. & Steele, W.G., *Experimentation and Uncertainty Analysis for Engineers*, second ed., John Wiley & Sons Inc., New York, 1999.
[10] Churchill, S.W. & Ozoe, H., Correlations for forced convection with uniform heating in flow over a plate and in developing and fully developed flow in a tube. *Journal of Heat Transfer*, **95**, pp. 78–84, 1973.

Section 3
Advances in computational methods

LBM mesoscale modelling of porous media

A. C. M. Sousa[1,2] & A. Nabovati[1]
[1]*Department of Mechanical Engineering, University of New Brunswick, Fredericton, NB, Canada*
[2]*Department of Mechanical Engineering, University of Aveiro, Aveiro, Portugal*

Abstract

Permeability is one of the most important bulk properties for the characterization of fluid flow in porous media. However, despite all the considerable body of research work over the past years using experimental, analytical, and numerical approaches, its determination is still a challenge. The methodologies, which have been used to measure, calculate and predict the permeability of different types of porous media, in general, tend to suffer from various levels of limitations in their applicability and, moreover, no general correlation for the permeability is available. Among the different predictive methods for the permeability, numerical pore level fluid flow analyses have been receiving increasing attention in the recent years, due to its robustness and flexibility. In this approach, the viscous fluid flow is directly simulated in the pores of the porous medium with no further modelling required. A simple representation of the pore structure can be in the form of the ordered and random packings of spheres, cylinders or square obstacles. In the present paper, the main objective is to introduce the lattice Boltzmann method (LBM) as a powerful tool for the mesoscopic pore level fluid flow simulation in porous media; two and three-dimensional case studies are presented to demonstrate the capabilities of the mesoscale modelling for porous media fluid flow problems using LBM. To demonstrate an approximation to a reconstructed medium, the fluid flow simulation in a 2D random arrangement of square obstacles with different aspect ratios is presented. Results of the three-dimensional simulations of the fluid flow in ordered packings of spheres are also reported; the results are in excellent agreement with the available analytical correlation for this configuration.
Keywords: pore level analysis, permeability prediction, LBM.

WIT Transactions on Engineering Sciences, Vol 61, © 2008 WIT Press
www.witpress.com, ISSN 1743-3533 (on-line)
doi:10.2495/HT080061

1 Introductions

The objective of the present paper is to review the lattice Boltzmann method (LBM), as a relatively new particle based method for the viscous fluid flow simulation, and to examine its performance when applied to the pore level simulation of fluid flow in porous media.

The investigation of the fluid flow in porous media at pore level enables the researchers to calculate the permeability and other bulk properties of porous media without any further modelling. The importance of the permeability is due to its role in any macroscopic study of fluid flow in porous media; either using the simplest possible model – the Darcy equation [1] or more recent and also more complicated, models (e.g. [2]); it is the permeability that characterizes the medium and it needs to be specified prior to any macro scale modelling. Determination of the permeability is a challenge and despite the considerable body of research work, almost all of the proposed correlations tend to suffer from various levels of restricted applicability; therefore, no general correlation for the permeability is available.

Literature dealing with the permeability prediction is vast and a considerable number of experimental, analytical and numerical studies have proposed different avenues and methodologies to predict the permeability.

The experimental approach has been popular among the researchers for years and still is one of the widely used methods to study the porous media flow (e.g. [3–9]). Jackson and James [10] and Skartsis *et al.* [11] have collected a comprehensive bibliography of the experimental and numerical research work conducted in this field.

Most of the analytical efforts to predict the permeability consider the porous media to be constructed as ordered arrangements of objects of simple geometrical shape, and the flow around them is described by a particular mathematical formulation. Emersleben in 1925, as reported in [11], was the first one who studied the flow parallel to square arrays of cylinders by solving the Navier Stokes equations. Larson and Higdon [12, 13] numerically studied the axial and transverse flow between square and hexagonal arrays of cylinders using the boundary integral method. Keller [14], Sangani and Acrivos [15], Drummond and Tahir [16], Higdon and Ford [17] have also considered the porous media to be in the form of ordered packing of cylinders with different arrangements.

Flow in different three-dimensional lattices of spheres was also analytically solved by Hasimoto [18] and Sangani and Acrivos [19]. The main shortcoming related to these studies is the lack of generality of the proposed correlations due to the simplifications introduced in the equations to make them amenable to the particular analytical method used.

An advanced method for the permeability prediction deals with the numerical simulation of the fluid flow in the inter-grain regions and inside the pores of the porous medium by solving either the Navier Stokes equations or their equivalent forms, such as, the Lattice Boltzmann equation. Then, the permeability of the medium can be calculated using the Darcy law [1] by averaging the local velocities over the flow, namely:

$$\langle \vec{u} \rangle = -\frac{\mathbf{K}}{\mu}\frac{dp}{d\vec{x}} \tag{1}$$

where $\langle \vec{u} \rangle$, $\mathbf{K}$, μ and $dp/d\vec{x}$ are the volume averaged velocity, permeability tensor, fluid viscosity and the applied pressure gradient.

The very first requirement in pore level analysis is to determine the grain structure and the pore geometry of the porous medium. The computed tomography (CT) imaging technique has had some success in the determination of pore structures [20, 21]; but, in general, the application of experimental imaging methods is costly and has its own limitations. An alternate approach is to reconstruct the porous medium by arrangements of the geometrical volumes, either orderly or randomly. Due to its simplicity and ability to reconstruct media similar to the actual structures, this method has been popular among the researchers in this field.

Numerically solving the fluid flow in the inter-grain voids, removes the limitation of considering simple geometric shapes for the grains; however, it still demands very high computational resources to resolve the inter-grain structure accurately, which for years had limited the applicability of the numerical methods to very simple geometries [22, 23]. In recent years, with the development of advanced numerical methods and increased availability of computational resources, more complex pore level geometries could be implemented. A particular difficulty related to the numerical pore level flow simulation approach is the limited applicability of the Navier Stokes equations to the micro flows, when local high velocities occur at the narrow pores, which is more likely to happen in the low porosity structures with random pore structure. The treatment of complex solid wall boundary conditions is also a cumbersome task when Navier Stokes equations are used. To overcome these two potential shortcomings of the pore level analysis, some novel methods for the fluid flow simulation have been proposed, namely: the Lattice Boltzmann Method [24, 25] and the Smoothed Particle Hydrodynamic [26, 27]. The main unresolved difficulty associated with the SPH method is the exponential computational cost with the increasing number of particles, which nearly precludes performing physically meaningful three-dimensional simulations. Similar problem was encountered with the early versions of the LBM introduced first by McNamara and Zanetti [28], but soon after it was removed by Qian *et al.* [29] and since then the LBM has been applied successfully to different problems of flow in porous media [30–34].

The present work addresses the LBM pore level fluid flow simulation in porous media - one of the earliest and most important fields of research to which the LBM has been successfully applied. Simulation results for the flow in two and three-dimensional reconstructed porous media are presented and the predicted permeability is compared against the available correlations to demonstrate the LBM capabilities.

2 Lattice Boltzmann method

Following the notation of Geller *et al.* [35], the form of the lattice Boltzmann equation (LBE) used in the present study can be regarded as the first order explicit upwind finite difference discretization of the discrete Boltzmann equation, as follows:

$$f_i(\vec{x}+\vec{e}_i\delta t, t+\delta t)-f_i(\vec{x},t)=\Omega_i(f_i(\vec{x},t)) \qquad i\in\{0,b\} \tag{2}$$

where f_i is the particle distribution function, $\vec{x}$ and t are the spatial coordinate and time, b+1 is the number of the discrete lattice speeds (in the present work, nine-speed lattice in 2D and 15-speed lattice in 3D simulations were employed), and Ω_i is the collision operator, which for the single relaxation time LBE is defined as follows:

$$\Omega_i=-\frac{\delta t}{\tau}\left(f_i(\vec{x},t)-f_i^{eq}(\vec{x},t)\right) \qquad i\in\{0,b\} \tag{3}$$

where τ is the lattice relaxation time and δt is the lattice time step, related to the lattice length scale, δx, as $c=\delta x/\delta t$; in most of the LBM simulations on a uniform lattice, including the present work, δx and δt are considered to be equal to one for simplicity. The equilibrium distribution function is the solution of the Boltzmann equation in the absence of any external force and zero spatial and temporal gradients.

Depending on the type of the lattice and number of the speeds, the particle distribution function has b+1 components, in the form of

$$f_i(\vec{x},t)=[f_0(\vec{x},t), f_1(\vec{x},t),\ldots, f_b(\vec{x},t)]^{\mathrm{T}} \tag{4}$$

Having calculated the distribution function on each lattice site using eqn (2), the average flow hydrodynamic parameters can be calculated as follows:

$$\rho=\sum_{i=0}^{b} f_i=\sum_{i=0}^{b} f_i^{eq} \tag{5}$$

$$\rho\vec{u}=\sum_{i=0}^{b} f_i\vec{e}_i=\sum_{i=0}^{b} f_i^{eq}\vec{e}_i \tag{6}$$

The kinematic viscosity of the fluid is calculated as a function of the relaxation time and the lattice time step as:

$$\nu=\frac{c^2}{3}\left(\tau-\frac{\delta t}{2}\right) \tag{7}$$

and the relation between the lattice sound speed and the pressure is as follows

$$p=\rho c_s^2 \tag{8}$$

where c_s is the lattice sound speed and is equal to $c/\sqrt{3}$. One of the main advantages of using the LBM is the easy implementation of the solid wall boundary condition, which in the present study was implemented by using the simple bounceback method [24] for both the two and three-dimensional simulations. The applied pressure gradient was treated as a uniform body force, which was added to the right hand side of eqn (2) at each time step on all the fluid nodes.

3 Results

A simple representation of the pore structure can be in the form of the ordered or random packings of spheres, cylinders or square obstacles. As a first approximation for a reconstructed medium, the fluid flow in 2D random arrangements of square obstacles of different aspect ratio was simulated by the present authors [34]. Predictions of the permeability were conducted and they were in good agreement with the available correlations. The effect of the obstacles' aspect ratio on the permeability was also studied.

For the three-dimensional study of the pore level flow, the packing of spheres is the most widely used structure; the reason is that the analytical flow solution around a sphere has been available in the literature for some time; also the packing of spheres resembles many of the real world porous media structures. Sangani and Acrivos [19] reported analytical correlations for the permeability of three different arrangements of spheres of constant radius – Simple Cubic (SC), Body Centred Cubic (BCC), and Face Centred Cubic (FCC) – in terms of series expansion solution of Stokes flow, as follows:

$$K = \frac{L^3}{6\beta\pi a C_d} \tag{9}$$

where K, a and L are the permeability, sphere radius and unit cube side length; β is the number of spheres in a cube of size L^3, which is equal to 1, 2 and 4 for the SC, BCC and FCC arrangements, respectively. C_d is calculated as follows:

$$C_d = \sum_{n=0}^{30} \alpha_n \chi^n \tag{10}$$

where the terms α_n are the series coefficients tabulated for each arrangement [19] and χ is a function of the solid volume fraction of the structure, c, as follows:

$$\chi = \left(c/c_{\max}\right)^{1/3} \tag{11}$$

$c_{\max}$ is the maximum solid fraction for each arrangement, determined by the maximum sphere radius in a cube of side length of L for which spheres are in contact without overlapping. $c_{\max}$ is equal to $\pi/6$, $\sqrt{3}\pi/8$ and $\sqrt{2}\pi/6$ for the SC, BCC and FCC structures, respectively.

Figure 1 shows the velocity magnitude contours and the velocity vectors in a BCC structure; cube side length (L) is 51 lattice units and the spheres' radius (a) is 16 lattice units, which yields a porosity of 0.87.

Figure 2 presents the comparison between the predicted permeability and the correlation of Sangani and Acrivos [19] for three different arrangements of spheres. As it can be seen, the agreement between the predicted values by the numerical simulation and the analytical correlation is excellent. For the low porosities, as was stated by Sangani and Acrivos [19], the series did not converge and the analytical solution is not accurate; which justifies the discrepancy between the numerical simulation and the analytical correlation.

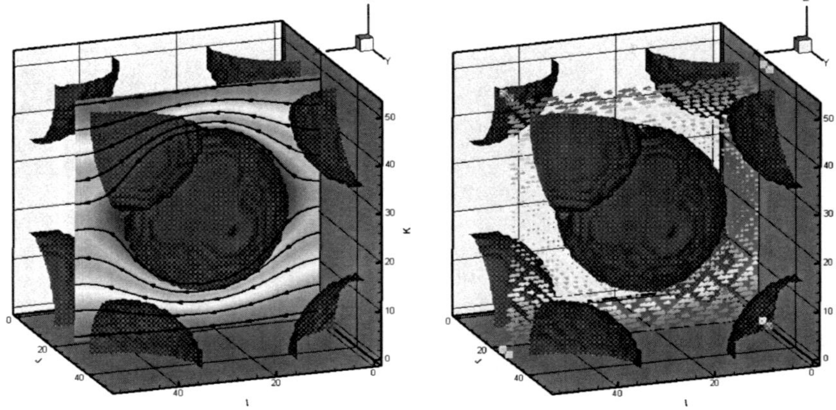

Figure 1: Velocity magnitude contours and the velocity vectors in a BCC structure; L=51 and a = 6 lattice units, which yields a porosity of 0.87.

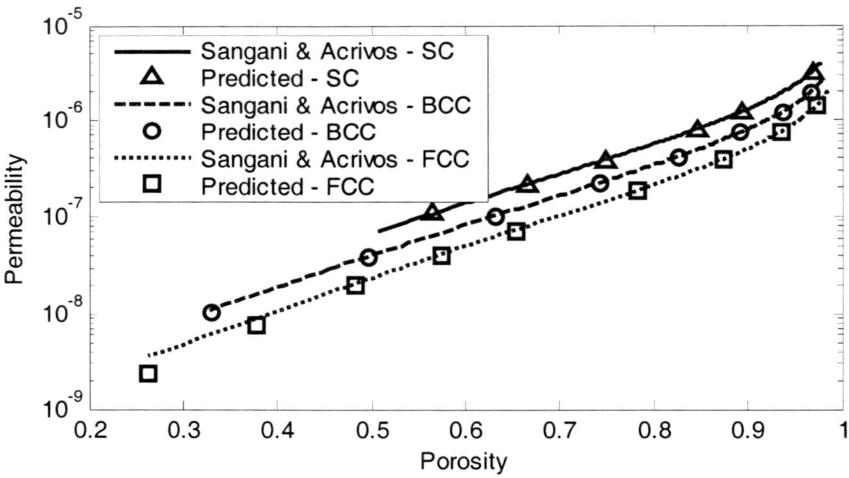

Figure 2: Comparison of the predicted permeability against the analytical correlation of Sangani and Acrivos [19] for three different arrangements of spheres, namely: SC, BCC, and FCC.

To change the porosity of each arrangement, size of the representative volume, L, was kept constant and the porosity was changed by varying the spheres' radius. As it is shown in figure 3, normalizing the predicted permeability by the square of the spheres' radius, a, for a constant value of L, causes the predicted values of the permeability for all the three different structures fall on the same curve, and eliminates the effect of the sphere arrangement when plotted as a function of the medium porosity. The normalized predicted permeability can be correlated as a function of the porosity, φ, as follows:

$$\frac{K}{a^2} = \exp\left(a_3\Phi^3 + a_2\Phi^2 + a_1\Phi + a_0\right) \tag{12}$$

$$\Phi = \ln\left(\frac{\varphi}{1-\varphi}\right) \tag{13}$$

where the fitting parameters, a_3, a_2, a_1, a_0 are calculated to be equal to 0.0118, -0.2093, 2.2648, and -4.6254, respectively. The correlation above is valid for all the three arrangements, namely SC, BCC, FCC, and for the entire range of the accessible porosity.

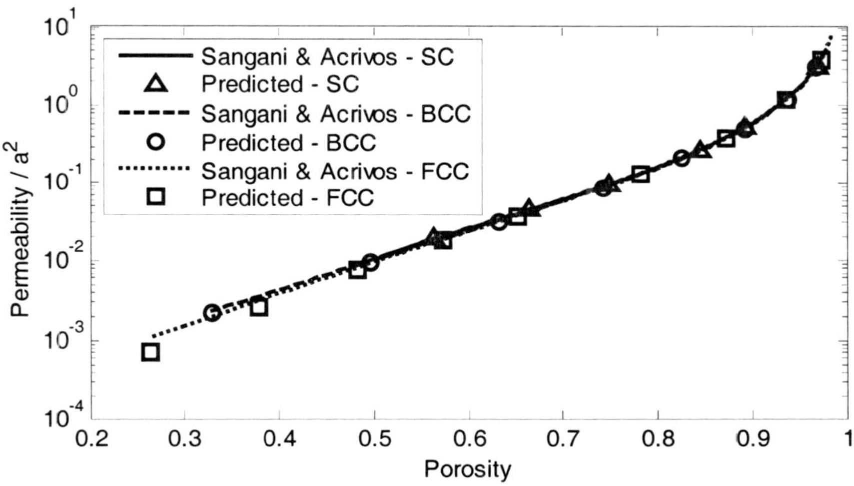

Figure 3: Normalized predicted permeability by the sphere radius versus porosity for three different arrangements of spheres, namely SC, BCC, and FCC.

4 Conclusion

The mesoscopic nature of the lattice Boltzmann method along with the implementation of the solid wall boundary condition make this approach a powerful tool for the pore level fluid flow simulation in porous media. Due to its mesoscopic nature, the LBM enables its use in both the macroscopic and microscopic flow simulations, where the conventional CFD methods based on the Navier Stokes equations require some modifications. Using the LBM for the pore level fluid flow simulations in porous media requires no further modelling and the bulk properties of the porous can be calculated as the outcome of the simulation.

The demonstration of the LBM capabilities was first performed in a previous publication, where the fluid flow was simulated in a two dimensional random arrangement of square obstacles and the effect of the aspect ratio on the permeability was studied. In the present work, to further demonstrate and test LBM, three-dimensional simulations were performed for the fluid flow in a three-dimensional ordered packing of spheres with three different arrangements, namely: SC, BCC, and FCC. Keeping the size of the representative volume constant, the porosity of the reconstructed medium was changed by varying the spheres' radius. The predicted permeability was in excellent agreement with the available analytical correlation for the entire range of the accessible porosity. Normalizing the permeability by the square of the spheres' radius causes all the predicted values to collapse on a single curve and eliminates the effect of the sphere arrangement on the permeability. A correlation for the normalized permeability as a function of the porosity was proposed, which is valid for the entire range of the porosity and the all the three arrangements.

Acknowledgements

The authors acknowledge the support received from NSERC (Natural Sciences and Engineering Research Council of Canada) Discovery Grant 12875 (ACMS) and from the Foundation for Science and Technology (FCT, Portugal) through the research grant POCTI/EME/59728/2004 (ACMS).

References

[1] Kaviany, M., *Principles of Heat Transfer in Porous Media*, Springer – Verlag: New York, 1991.

[2] Vafai, K. & Thiyagaraja, R., Analysis of flow and heat transfer at the interface region of a porous medium", International Journal of Heat and Mass Transfer, **30(7)**, pp. 1391–1405, 1987.

[3] Tomadakis, M.M. & Robertson, T.J., Viscous permeability of random fiber structures: Comparison of electrical and diffusional estimates with experimental and analytical results, Journal of Composite Materials, **9**, pp. 163–188, 2005.

[4] Loosveldt, H., Lafhaj, Z. & Skoczylas, F., Experimental study of gas and liquid permeability of a mortar, Cement and Concrete Research, **32(9)**, pp. 1357–1363, 2002.

[5] Seguin, D., Montillet, A. & Comiti, J., Experimental characterization of flow regimes in various porous media - I: Limit of laminar flow regime, Chemical Engineering Science, **53(21)**, pp. 3751–3761, 1998.

[6] Rahli, O., Tadrist, L. & Miscevic, M., Experimental analysis of fibrous porous media permeability, AICHE Journal, **42(12)**, pp. 3547–3549, 1996.
[7] Rashidi, M., Peurrung, L., Tompson, A.F.B. & Kulp, T.J., Experimental analysis of pore-scale flow and transport in porous media, Advances in Water Resources, **19(3)**, pp. 163–180, 1996.
[8] Comiti, J. & Renaud, M., A new model for determining mean structure parameters of fixed-beds from pressure-drop measurements - application to beds packed with parallelepipedal particles, Chemical Engineering Science, **44(7)**, pp. 1539–1545, 1989.
[9] Williams, J.G., Morris C.E.M. & Ennis, B.C., Liquid Flow through Aligned Fiber Beds, Polymer Engineering and Science, **14(6)**, pp. 413–419, 1974.
[10] Jackson, G.W. & James, D.F., The Permeability of Fibrous Porous-Media, Canadian Journal of Chemical Engineering, **64(3)**, pp. 364–374, 1986.
[11] Skartsis, L., Khomami, B. & Kardos, J.L., Resin Flow Through Fiber Beds During Composite Manufacturing Processes. Part II: Numerical and Experimental Studies of Newtonian Flow Through Ideal and Actual Fiber Beds, Polymer Engineering and Science, **32(4)**, pp. 231–239, 1992.
[12] Larson, R.E. & Higdon, J.J.L., Microscopic Flow near the Surface of Two-Dimensional Porous-Media. Part 1. Axial Flow, Journal of Fluid Mechanics, **166**, pp. 449–472, 1986.
[13] Larson, R.E. & Higdon, J.J.L., Microscopic Flow near the Surface of Two-Dimensional Porous-Media. Part 2. Transverse Flow, Journal of Fluid Mechanics, **178**, pp. 119–136, 1987.
[14] Keller, J.B., Viscous flow through a grating or lattice of cylinders, Journal of Fluid Mechanics, **18**, pp. 94–96, 1964.
[15] Sangani, A.S. & Acrivos, A., Slow flow past periodic arrays of cylinders with application to heat transfer, International Journal of Multiphase Flow, **8(3)**, pp. 193–206, 1982.
[16] Drummond, J.E. & Tahir, M.I., Laminar viscous flow through regular arrays of parallel solid cylinders, International Journal of Multiphase Flow, **10(5)**, pp. 515–540, 1984.
[17] Higdon, J.J.L. & Ford, G.D., Permeability of three-dimensional models of fibrous porous media, Journal of Fluid Mechanics, **308**, pp. 341–361, 1996.
[18] Hasimoto, H., On the periodic fundamental solutions of the stokes equations and their application to viscous flow past a cubic array of spheres, Journal of Fluid Mechanics, **5**, pp. 317–328, 1959.
[19] Sangani, A.S. & Acrivos, A., Slow flow through a periodic array of spheres, International Journal of Multiphase Flow, **8(4)**, pp. 343–360, 1982.
[20] Heijs, A.W.J. & Lowe, C.P., Numerical evaluation of the permeability and the Kozeny constant for two types of porous media, Physical Review E, **51**, pp. 4346–4353, 1995.
[21] Hazlett, R.D., Simulation of Capillary-Dominated Displacements in Microtomographic Images of Reservoir Rocks, Transport in Porous Media, **20**, pp. 21–35, 1995.

[22] Sahraoui, M. & Kaviany, M., Slip and no slip velocity boundary conditions at interface of porous, plain media, International Journal of Heat and Mass Transfer, **35(4)**, pp. 927–943, 1992.

[23] Nakayama, A., Kuwahara, F., Umemoto, T. & Hayashi, T., Heat and Fluid Flow Within an anisotropic porous medium, Journal of Heat Transfer, **124**, pp. 746–753, 2002.

[24] Succi, S., *The Lattice Boltzmann Equation - For Fluid Dynamics and Beyond*, Oxford University Press: Oxford, 2001.

[25] Chen, S. & Doolen, G.D., Lattice Boltzmann method for fluid flows, Annual Review of Fluid Mechanics, **30**, pp. 329–364, 1998.

[26] Jiang, F., Oliveira, M.S.A. & Sousa, A.C.M., Mesoscale SPH modeling of fluid flow in isotropic porous media, Computer Physics Communications, **176 (7)**, pp. 471–480, 2007.

[27] Jiang, F. & Sousa, A.C.M., Smoothed Particle Hydrodynamics Modeling of Transverse Flow in Randomly Aligned Fibrous Porous Media, Transport in Porous Media, DOI: 10.1007/s11242-008-9206-z, available online since February 2008.

[28] McNamara, G.R. & Zanetti, G., Use of the Boltzmann Equation to Simulate Lattice-Gas Automata, Physical Review Letters, **61(20)**, pp. 2332–2335, 1988.

[29] Qian, Y.H., d'Humieres, D. & Lallemand, P., Lattice BGK model for Navier-Stokes equation, Europhysics Letters, **17(6BIS)**, pp. 479–484, 1992.

[30] Succi, S., Foti, E. & Higuera, F., Three-dimensional flows in Complex Geometries with the Lattice Boltzmann Method, Europhysics Letters, **10(5)**, pp. 433–438, 1989.

[31] Cancelliere, A., Chang, C., Foti, E., Rothman, D.H. & Succi, S., The permeability of a random medium: Comparison of simulation with theory, Physics of Fluids A, **2(12)**, pp. 2085–2088, 1990.

[32] Koponen, A., Kataja, M. & Timonen, J., Permeability and effective porosity of porous media, Physical Review E, **56(3)**, pp. 3319–3325, 1997.

[33] Nabovati, A. & Sousa, A.C.M., Fluid Flow Simulation at the Open-Porous interface Using Lattice Boltzmann Method, International Journal for Numerical Methods in Fluids, 56(8), pp. 1449 – 1456, 2008.

[34] Nabovati, A. & Sousa, A.C.M., Fluid Flow Simulation in Random Porous Media at Pore Level using the Lattice Boltzmann Method, Journal of Engineering Science and Technology, **2(3)**, pp. 226– 237, 2007.

[35] Geller, S., Krafczyk, M., Tolke, J., Turek, S. & Hron, J., Benchmark computations based on lattice-Boltzmann, finite element and finite volume methods for laminar flows, Computers and Fluids, **35**, pp. 888–897, 2006.

Chemical reacting transport phenomena and multiscale models for SOFCs

M. Andersson, J. Yuan & B. Sundén
Department of Energy Sciences, Lund University, Sweden

Abstract

Electrochemical reactions at the anode triple phase boundary (TPB) proceed on the basis of the fuel concentration, which depends on transport processes within the porous anode and the heterogeneous reforming chemistry. Microscale modeling is needed to describe these interactions with an acceptable accuracy. The aim of this article is to investigate if it is possible to use a multiscale approach to model solid oxide fuel cells (SOFCs) and combine the accuracy at microscale with for example the calculation speed at macroscale to design SOFCs, based on a clear understanding of transport phenomena and functional requirements. A literature review is made to find out what methods can be used to model SOFCs and also to sort these models after length scale. Couplings between different methods and length scales, i.e., multiscale modeling, are outlined. The SOFC microscale model corresponds in many cases to the atom or molecular level, such as Lattice Bolzmann Method, Density Functional Theory, Molecular Dynamics, Dusty Gas Model, Ficks Model and Stefan-Maxwell Model. SOFC modeling in the mesoscale can be done with Kinetic Monte Carlo. Macroscale models match to the global flow field. Finite Element Method and Finite Volume Method are used to model SOFCs in the macroscale. Multiscale modeling is a promising tool for fuel cell research. COMSOL Multiphysics, based on the Finite Element Method as well as FLUENT, based on the Finite Volume Method, can be used to couple different physical models at different scales. Multiscale modeling increases the understanding for detailed transport phenomena, and can be used to make a correct decision on the specific design and control of operating conditions. It is expected that the development- and production cost will decrease as the understanding of complex phenomena increases.

Keywords: SOFC, reactions, transport phenomena, multiscale modelling.

WIT Transactions on Engineering Sciences, Vol 61, © 2008 WIT Press
www.witpress.com, ISSN 1743-3533 (on-line)
doi:10.2495/HT080071

1 Introduction

Fuel cells convert the free energy of a chemical reactant to electrical energy and heat. This is different from a conventional thermal power plant, where the fuel is oxidised in a combustion process combined with a conversion process (thermal-mechanical-electrical energy), that takes place after the combustion. Fuel cells can have conversion efficiencies higher than heat engines, since there is no Carnot cycle efficiency limitation [1]. Anode, cathode and electrolyte are the basic building blocks for a fuel cell, their structures decide the efficiency. If pure hydrogen is used no pollution of air and environment occurs, the output from the fuel cell is electricity, heat and water [2].

The reality can often be described at different levels of scale; both length scale and time scale can be varied. This concept can be applied both quantitatively and qualitatively. Modeling a system on a coarser level has two advantages: the modeling is faster and unnecessary details are disregarded [3]. However, at each level of homogenization or scale-up, the risk of losing the key structural information increases [4].

The challenge for the future is to use multiscale modeling to design an optimized nano-structured fuel cell, which is based on functional requirements of a SOFC system [5]. Multiscale modeling is used to understand detailed transport phenomena, to specify the design and control the operating conditions. Combination of micro- and macroscopic analyses provides an engineer a multiscale design tool to understand the detailed transport phenomena and to make a good SOFC design [6]. Fuel cell operation depends on thermal, electrochemical, mechanical and chemical phenomena. A multiscale system-to-science approach is enabled with a multi-physics balance. Cost is reduced, durability is improved and technology development is accelerated [5].

2 SOFC features

A fuel cell consists of two electrodes, an anode for fuel and a cathode for oxidant. The electrodes are separated by the electrolyte and connected into an electrically conducting circuit. A gas with fuel or oxidant is transported to the electrode, which
should be permeable via a porous structure. Cells are organised together into stacks [8]. The structure and function (fig. 1) of the different parts of an SOFC are described in this chapter.

An SOFC anode is a mixed ionic-electronic conductor and it consist of either Ni/YSZ cermet or Ni/Ce(Sm, Gd)O_2. The electronic conductivity is associated with a conductive percolation path and the ionic conductivity is related to the oxygen vacancies in YSZ or Ni/Ce(Sm, Gd)O_2. The anode material contains about 30% wt. of Ni, which works as a catalyst for the oxidation of fuel. An SOFC cathode works in oxygen or air atmosphere. It consist of $Ln_{1-x}A_xMO_3$, where Ln = lanthanides, A = alkali-earth metals and M = Mn, Fe, Co or Ni, such materials are mixed ionic-electronic conductors. The electronic conductivity results from a mixed valency of M^{3+}/M^{4+} and the ionic conductivity steams from the deviation from stoichiometry in oxygen vacancies. Electrolytes are in contact with the oxidizing atmosphere at the cathode side and with the highly reductive atmosphere at the anode side. Required electronic conductivity is about 10^{-1} S cm^{-1} [9]. The nonporous ceramic electrolyte becomes conductive to oxygen ions, O^{2-}, but non-conductive to electrons at the working temperature for SOFCs, 600-1000°C [10].

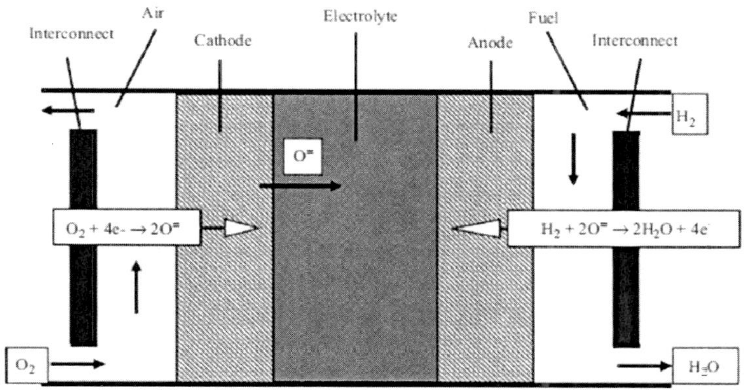

Figure 1: Planar SOFC structure [7].

SOFCs can work with a variety of fuels e.g., hydrogen, carbon monoxide, methane and combinations of these [11]. Oxygen is reduced in the cathode, eqn. (1). The electrons are received from outside the cell and the oxygen ions are delivered to the electrolyte. The electrochemical reactions, eqns. (2)-(3), take place in the anode close to the electrolyte. Methane needs to be reformed, eqn. (4), before the electrochemical reactions [12]. Carbon monoxide can be oxidized in the electrochemical reaction, eqn. (3), but can also react with water eqn. (5). The reactions described here are the overall reactions, more detailed reaction mechanisms can be found in [12–14].

The gas flow in the air and fuel channels can for most cases be considered as one-dimensional and the gas flow is highly viscous and a fully developed velocity profile is established close to the channel entrance. The diffusion in an SOFC is for light gas species around 1 cm/s, compared to convective velocities of the order of 10 to 100 cm/s [11]. Transport in the porous electrodes occurs both in gas- and in solid phase, integrated with the electrochemical reactions. The electrodes are porous and mass transfer is dominated by gas diffusion [15]. The electrolyte has two functions: to transport oxide ions form the electrolyte to the anode and to block electron flow from the anode to the cathode inside the cell [12]. The flow of electronic charge through an outside circuit balances the flow of ionic charge through the electrolyte and electrical power is produced [16]. The interconnect can be assumed to be impermeable for planar SOFC material. Electron transport needs to be considered since current from the SOFC is collected [7].

$$O_2 + 4e^- \Leftrightarrow 2O^{2-} \tag{1}$$

$$H_2 + O^{2-} \Leftrightarrow H_2O + 2e^- \quad (2)$$

$$CO + O^{2-} \Leftrightarrow CO_2 + 2e^- \quad (3)$$

$$CH_4 + H_2O \Leftrightarrow 3H_2 + CO \quad (4)$$

$$CO + H_2O \Leftrightarrow H_2 + CO_2 \quad (5)$$

The amount of fuel gases transported to the active surface for electrochemical reactions is governed by different parameters, such as porous microstructure, gas consumption, pressure gradient between the fuel flow duct and the porous anode and inlet conditions [17]. The gas molecules diffuse to the pore walls, where electrocatalytic reaction takes place. The supply of reactant is the rate limiting step, since the gas molecule diffusion coefficient is much smaller than for ions. The charge transfer chemistry at the interface between the electrolyte and the anode proceeds on the basis of the hydrogen concentration. The hydrogen concentration depends on the transport within the porous anode and the heterogeneous reforming chemistry. The concentration of the fuel gases, CH_4, CO and H_2, decreases along the length of the fuel channel while the concentration increases for H_2O and CO_2. As a result the current density decreases along the fuel channel [11]. The most active area for reaction inside a SOFC is the TPB. However, not all surface reactions occur there. Depending on the material design, other parts could be as important. The adsorption-desorption reaction may expand onto the electrode surface if the surface diffusion is fast compared to the surface reaction. The oxygen transport path can expand into the electrode/electrolyte boundary if the diffusion along the electrode/electrolyte boundary or the bulk diffusion electrolyte is fast [18].

3 Models

Before designing and constructing a model, it is important to specify what one wants to know, how accurate and why. The choice of computational methods must come from a clear understanding of both the information being computed and the chemical system. It is also needed to be aware of what approximations being made and which one being significant [19].

SOFCs can be examined from different points of view; as an electrochemical generator in a viewpoint of electrochemical modeling at continuum level, as a heat and mass exchanger in a viewpoint of fluid dynamics and transport phenomena, as a chemical reactor in viewpoints of chemical reactions depending on fuel composition and heat effects associated with their electrochemical conversion [7].

An SOFC can be described by different length scales: system scale (~10^2 m), component scale (~10^1 m), material aspect at the fuel cell/constituent (~10^{-2} m), flow/diffusion morphologies (~10^{-3} m), material structure/interface (~10^{-6} m), and functional material levels (~10^{-9} m) [5].

Not only proper length scales are needed to describe various parts of a SOFC, also different time scales need to be considered. Cell charging and cathode gas thermal diffusion in 10^{-3} s, convective transport in 10^{-1}s, cell heating and anode streamwise thermal diffusion in 10^3 s and cathode streamwise thermal diffusion in 10^4 s [21]. The relation between time- and length scales can be seen in fig. 2.

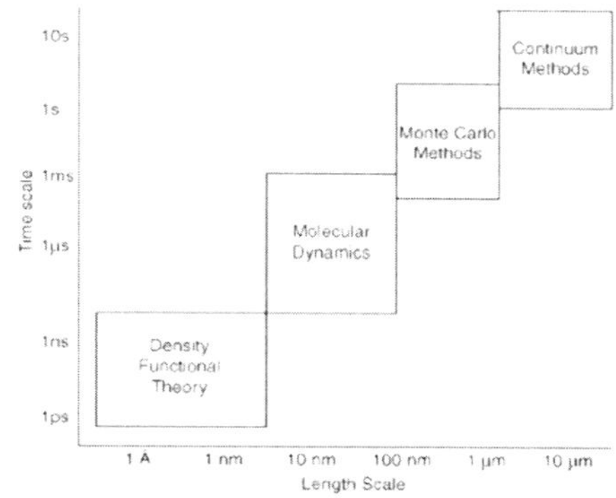

Figure 2: Characteristic time and length scales for various methods [20].

3.1 Micro-, meso- and macro scale approaches

Research of natural phenomena is based on different levels of scale; micro-, meso- and macroscales. The microscale model corresponds in many cases to the atom or molecular level when thermo- or fluid dynamics is studied. The microscale does not need to be as small as the size of the molecules. A mesoscale model corresponds to a larger scale than a particle but a smaller one than the facility or the global flow field. Macroscale models match to the global flow field. Microscale modeling is in general more related to theoretical knowledge compared to macroscale modeling that is more related to empirical data. Empirical factors in a macro-model could be based on the result from a micro- or meso model [22].

Instantaneous flow around individual moving particles can be calculated in microscale models. Flows corresponding to calculation cells, larger than particles but smaller than global flow field, are calculated with mesoscale models. Trajectories of individual particles are calculated with particle motion equations for micro scale modeling. The flow field is in mesoscale modeling divided into a number of small cells, but not as small as the particles size [22].

Table 1: Computational methods sorted after scale [20–33].

Microscale	Mesoscale	Macroscale
Density Functional Theory (DFT)	Kinetic Monte Carlo	Finite Element Method (FEM)
Quantum Chemistry (QC)	(KMC)	Finite Volume Method (FVM)
Lattice Bolzmann Method (LBM)	Brownian Dynamics	Finite Differnece Method (FDM)
Molecular Dynamics (MD)	Dissipative Particle	Spectral Methods (SM)
Mechanistic Models	Dynamics (DPD)	
Dusty Gas Model (DGM)		
Ficks Model (FM)		
Stefan-Maxwell Model (SMM)		

Different methods have been developed to describe different scales. Methods that have been used for SOFC modeling are listed in Table 1. The methods are divided into microscale, mesoscale and macroscale. Examples of use for the different methods are described in the following chapters.

3.1.1 Modeling at the micro scale

Frayret *et al* [25] simulated microscopic aspects of oxygen diffusion in the ceria-based material in the ionic conductor with Density Functional Theory (DFT). This methodology is a good tool to study the connection between dopant ionic radius and diffusion at the atomistic scale. DFT is an "ab inito" method where the material properties are described by solutions of the Schrödinger equations. DFT models have a characteristic length scale of Å – nm and a time scale of ps – ns.

Cheng *et al* [26] simulated the oxygen ion-hoping phenomenon inside a YSZ electrolyte with Molecular Dynamics (MD). MD can be used to model grain boundary structure, specific heat capacity and molecular structure. Systems up to 10^5 atoms and a time scale on the order of ten ns can be modeled [6].

The Lattice Bolzmann Method (LBM) is used to model mass transport of gases inside the porous anode of an SOFC. The porous structure is based on SEM images, which are converted to digital form. Advantages of the LBM model are that a detailed analysis of mass transfer can be carried out for the actual anode microstructure; this means that tortuosity is not used as a fitting parameter. LBM approach is according to an investigation in [25] accurate enough to model concentration polarization, in 2D, in a SOFC, see Fig. 3(a), where the mole fraction distribution for H_2 is shown. By changing the amount of void spaces present in the solid matrix the porosity in the LBM is varied.

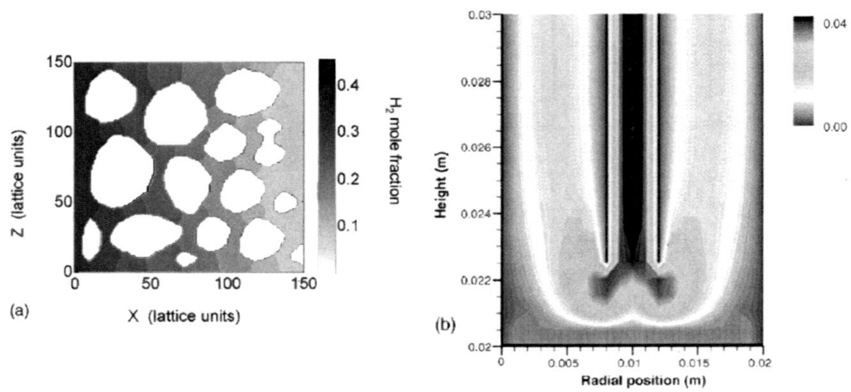

Figure 3: (a) Mole fraction distribution for H_2 is calculated with microscale LBM [25]. (b) Dusty gas model is coupled to FVM in FLUENT to calculate the velocity within the anode compartment for a SOFC with tubular design. A modified Butler-Volmer equation and multi-step chemical reactions are parts of the model [14].

Dusty Gas model (DGM), Ficks model (FM) and Stefan-Maxwell model (SMM) are developed in [26] in order to predict the concentration over potential inside an SOFC anode. DGM and FM consider molecular diffusion, Knudsen diffusion and the effect of a finite pressure gradient. The flux ratio in DGM depends on the square-root of the gas molecular weight, but it does not for FM. Explicit analytical expressions, which describe fluxes, can be used in FM. The SMM can be seen as a simpler model since it does not consider the Knudsen diffusion. DGM is the most appropriate model for H_2-H_2O and CO-CO_2 system. However, it is only required to use it when the operating current density is high. Ni *et al* [24,29] studied the effect of micro-structural grading on SOFC performance with a DGM, to analyse the coupled phenomena of mass transfer and electrochemical reactions in the SOFC electrodes.

3.1.2 Modeling at the meso scale

Modak and Lusk [6] applied Kinetic Monte Carlo (KMC) to simulate the open-circuit voltage and electrical double layer of a doped electrolyte. Discrete time increments of varying size are used to capture diffusion or adsorption in a single step. The physical property data generated by QC and MD can be utilized in the KMC model. Monte Carlo methods have a characteristic length scale of 100 nm – μm and a time scale of ms – s [23].

Huang *et al* [30] employed COMSOL Multhiphysics (FEM) to model the multhiphysics processes in the SOFC cathode-electrolyte interfaces considering the geometry and detailed distribution of the pores and the ionic conducting phase. The charge transfer rate, electrical conduction and ion conduction are governed on a modeling domain abstracted from actual materials encountered in the application.

3.1.3 Modeling at the macro scale

Cheng *et al* [26] used the Finite Element Methods (FEM) and the commercial software COMSOL Multiphysics to solve the flow equations for macroscopic transport phenomena. Navier-Stokes equations are used to describe the flow conditions in the air and fuel channels and Darcy law describes the flow conditions in the porous layer. FEM has a characteristic time scale of 1 s and above [23].

A clear relationship between underlying physical conditions and numerical algorithm have made Finite Volume Method (FVM) a popular method for commercial codes such as PHOENICS, FLUENT, CFX and STAR-CD [30]. Pasaogullari and Wang [33] as well as Autossier *et al* [34] used FLUENT to solve equations for momentum, mass, energy, multicomponent species and electrochemical kinetics for an SOFC. Hussain *et al* [35] employed the FVM to model the transport of multi-component species inside porous SOFC anodes.

3.2 Integration issues

Multiphysics modeling considers interactions, couplings, between two or more physical disciplines. Physical problems can often be described with a set of partial differential equations. The coupled partial differential equations can be solved simultaneously in potentially physical domains for each corresponding physical phenomena. Fuel cell operation depends on complex interaction between multi- physics such as multi-phase fluid flow, mass transport, heat transfer and electrochemical reactions [34]. Two basic integration approaches can be found: hierarchical methods, and hybrid and cocurrent method. The hierarchical modeling starts at higher resolution (smaller scale) and properties are extracted and used as input to the next level method. The hierarchical methods are today the most developed methods for multiscale modeling [23]. Three different methods are used for hybrid approaches to describe various regions of the material with the appropriate time and length scale resolutions. The hybrid methods that permit cocurrent simulations are promising for the future development, since only one calculation needs to be performed, however, it requires more computational power compared to hierarchical methods [23].

The particle size in SOFCs is in the sub-micron scale, and the TPBs are in nano-scale. The morphology and properties of these scales are important for the performance of the fuel cell, since they control how much of the Gibbs free energy being available for use. This means that the science at nano-scale is critical to the performance at a system-scale. A robust design and multi-scale analysis consider those nano-details as well as macro system level [5].

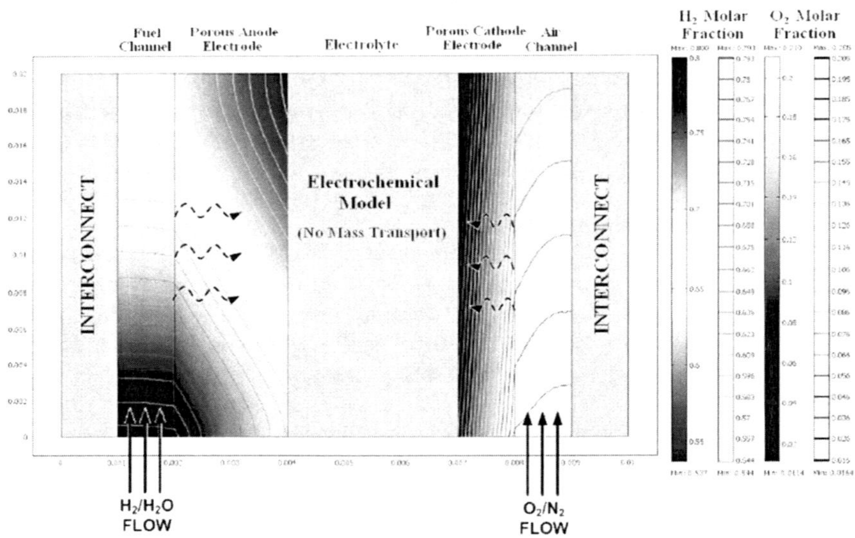

Figure 4: The concentration of O_2 and H_2 in a SOFC after a multiscale modeling, where Stefan-Maxwell model is used for the flow in air and fuel channels, Dusty-Gas model for transport in porous media, and Butler-Volmer equation for electrochemical phenomena among others [29].

Tseronis *et al* [31] developed a multidimensional concept, where the microscale DGM is used to describe mass transfer in porous media and FEM based COMSOL Multhiphysics is used for the numerical solution of the model equations. Figure 3(b) illustrates the velocity within the anode compartment. DGM is used for porous media transport, Butler-Volmer equation for electrochemicstry, a multi-step model for heterogeneous chemistry and FLUENT is used to couple the different physical descriptions together [14]. Figure 4 illustrates the concentration of H_2 and O_2 for a multiscale modeling approach in COMSOL Multiphysics, where Stefan-Maxwell model describes the flow in the air and fuel channels, Butler-Volmer equation describes electrochemical phenomena, ionic- and electronic potentials are described by ohmic phenomena and heat transfer is described by radiation, conduction and convection [29].

Cheng *et al* [26] and Pasaogullari and Wang [33] introduced multi-scale-concepts; however, they did not mention how the different scales interact with each other. It is frequently stated in literature that data or property constants can be obtained from smaller scale and then used in a model that is made in a larger scale. However, information about the construction of this coupling is rare [14,20–23].

4 Conclusions

SOFCs can be described at different length- and time scales. A challenge for the future is to further develop multiscale models for fuel cell designs. Multiscale models can provide a clear understanding of operating conditions, transport and reaction phenomena at the microscale connected to, e.g., conditions in the air- and fuel channels at the macroscale.

COMSOL Multiphysics and FLUENT can be used for modeling with multiscale approach. It is possible to couple different physical models, for example models at the microscale, describing transport phenomena inside an anode, to a macroscale model describing the entire fuel cell. It is important to specify what one wants to know and how accurate before designing a model.

Use of multiscale modeling in fuel cell research will lead to an increased power density and also to a lower cost for development and production, i.e., the commercialization of fuel cells will be promoted.

Acknowledgement

The Swedish Research Council (VR) supports the current research.

References

[1] Li P.-W., Schaefer L., Chyu M.K., Multiple Transport Processes in Solid Oxide Fuel Cells, *Transport Phenomena in Fuel Cells*, Sundén B. and Faghri M. (eds.), WIT Press, pp. 1–41, 2005

[2] Yuan J., Faghri M., Sundén B., On Heat and Mass Transfer Phenomena in PEMFC and SOFC and Modelling Approaches, *Transport Phenomena in Fuel Cells*, Sundén B. and Faghri M. (eds.), WIT Press, pp. 143–174, 2005

[3] Müller-Plathe F., Coarse-Graining in Polymer Simulation: From the Atomistic to the Mesoscopic Scale and Back, *ChemPhysChem*, 3, pp. 754–769, 2002

[4] Gates T., Odegard G., Frankland S., Clancy T., Computational materials: Multi-scale modeling and simulation of nanostructured materials, *Comp. Sci. Technol.*, 65, pp. 2416–2434, 2005

[5] Reifsnider K., Hunag X., Ju G., Solasi R., Multi-scale Modeling Approaches for Functional Nano-composite Materials, *J. Mater. Sci.*, 41, pp. 6751–6759, 2006

[6] Modak A.U., Lusk M.T., Kinetic Monte Carlo Simulation of a Solid-oxide Fuel Cell, *Solid State Ionics*, 176, pp. 1281–1291, 2005

[7] Kackac S., Pramuanjaroenkij A., Zhou X., A Review of Numerical Modeling of Solid Oxide Fuel Cells, *I. J. Hydrogen Energy*, 32, pp. 761–786, 2007

[8] Beale S.B., Numerical Models for Planar Solid Oxide Fuel Cells, *Transport Phenomena in Fuel Cells*, Sundén B. and Faghri M. (eds.), WIT Press, pp. 42–82, 2005

[9] Molenda J., Swierczek K., Zajac W., Functional materials for the IT-SOFC, *J. Power Sources*, 173, pp. 657–670, 2007

[10] Kim S., Lee J., Moon H., Hyun S., Miin J., Kim J., Lee H., Effects of Anode and Electrolyte Microstructures on Performance of Solid Oxide Fuel Cells, *J. Power Sources,* 169, pp. 265–270, 2007

[12] Zhu H., Kee R., Janardhanan V., Deutschmann O., Goodwin D., Modeling Elementary Heterogeneous Chemistry and Electrochemistry in Solid-Oxide Fuel Cells, *J. Electrochem. Soc.*, 152, pp. A2427–A2440, 2005

[13] Goodenough J.B. Huang Y., Alternative Anode Materials for Solid Oxide Fuel Cells, *J. Power Sources*, 173, pp. 1–10, 2007

[14] Bessler W.G., Gewies S., Vogler M., A New Framework for Physically Based Modeling of Solid Oxide Fuel Cells, *Electrochim. Acta*, 53, pp. 1782–1800, 2007

[15] Janardhanan V., Deutschmann O., CFD Analysis of a Solid Oxide Fuel Cell with Internal Reforming, *J. Power Sources*, 162, pp. 1192–1202, 2006

[16] Nam J. H., Jeon D. H., A Comprehensive Micro-scale Model for Transport and Reaction in Intermediate Temperature Solid Oxide Fuel Cells, *Electrochim. Acta*, 51, pp. 3446–3460, 2006

[17] Haile S.M. Fuel Cell Materials and Components, *Acta Mater.*, 51, pp. 1981–2000, 2003

[18] Yuan J., Ren F., Sunden B., Analysis of Chemically Reacting Transport Phenomena in an Anode Duct of Intermediate Temperature SOFCs, *J. Fuel Cell Sci. Technol.*, 3, pp. 687–701, 2006

[19] Kawada T., Sase M., Kudo M., Yashiro K., Sato K., Mizusaki J., Sakai N., Horita T., Yamaji K., Yokokawa H., Microscopic Observation of Oxygen Reaction Pathway on High Temperature Electrode Materials, *Solid State Ionics,* 177, pp. 3081–3086, 2006

[20] Young D., Computational Chemistry A practical guide for Applying Techniques to Real-World Problems, John Wiley & Sons, 2001

[22] Karakasidis T., Charitidis C. Multiscale modelling in nanomaterials science, *Mater. Sci. Eng.*, C27, pp. 1082–1089, 2007

[23] Kemm M., Dynamic Solid Oxide Fuel Cell Modelling for Non-steady State Simulation of System Applications, Doctoral thesis, Department of Energy Sciences, Lund University, 2006

[24] Ni, M., Leung M. K. H., Leung D.Y.C., Micro-Scale Modeling of Solid Oxide Fuel Cells with Micro-structurally Grades Electrodes, *J. Power Sources*, 168, pp. 369–378, 2007

[25] Frayret C., Villesuzanne A., Pouchard M., Matar S., Density Functional Theory Calculations on Microscopic Aspects of Oxygen Diffusion in Ceria-Based Materials, *I. J. Quantum Chemistry*, 101, pp. 826–839, 2005

[26] Cheng C.H., Chang Y.W., Hong C.W. Multiscale Parametric Studies on the Transport Phenomenon of a Solid Oxide Fuel Cell, *J. Fuel Cell Sci. Technol.*, 2, pp. 219–225, 2005

[27] Joshi A.S., Grew K.N., Peracchio A.A., Chiu W.K.S., Lattice Boltzmann Modeling of 2D Gas Transport in a Solid Oxide Fuel Cell Anode, *J. Power Sources,* 164, pp. 631–638, 2007

[28] Suwanwarangkul R., Croiset E., Fowler M.W., Douglas P.L., Entchev E., Douglas M.A., Performance Comparison of Fick's, Dusty-gas and Stefan-Maxwell Models to Predict the Concentration Overpotential of a SOFC Anode, *J. Power Sources*, 122, pp. 9–18, 2003

[29] Ni, M., Leung M. K. H., Leung D.Y.C., Micro-Scale Modeling of a Functionally Graded Ni-YSZ Anode, *Chem. Eng. Technol.*, pp. 287–592, 2007

[30] Huang W., Huang X., Reifsnider K., Meso-Scale Multiphysics Model of SOFC Cathode Processes, *COMSOL Users Conference Boston*, 2006

[31] Tseronis K., Kookos I., Theodoropoulos K., Modelling and Design of the Solid Oxide Fuel Cell Anode, *COMSOL Users Conference Birmingham*, 2006

[32] Versteeg H.K., Malalasekera W, An Introduction to Computational Fluid Dynamics, The Finite Volume Method, Pearson, 1995

[33] Pasaogullari U., Wang C.-Y., Computational Fluid Dynamics Modeling of Solid Oxide Fuel Cells, *Proceedings of SOFC-VIII,* Eds. S.C. Singhal and M. Dokiya, pp. 1403–1412, 2003

[34] Autissier N., Larrain D., Van Herle J., Favrat D., CFD Simulation Tool for Solid Oxide Fuel Cells, *J. Power Sources*, 131, pp. 131–319, 2004

[35] Hussein M. M., Li X., Dincer I., Mathematical Modeling of Transport Phenomena in Porous SOFC Anodes, *Int. J. Thermal Sciences*, 46, pp. 48–86, 2007

[36] Dennis, B.H., Han Z., Jin W., Wang B.P., Xu L., Aapro, T., Ptchelintsev, A., Reinikainen, T., Multi-Physics Simulation Strategies with Application to Fuel Cell Modeling, *7th. Int. Conf. on Thermal, Mechanical and Multiphysics Simulation and Experiments in Micro-Electronics and Micro-Systems,* EuroSimE, 2006

Heat transfer between shell and rigid body through the thin heat-conducting layer taking into account mechanical contact

V. V. Zozulya
Centro de Investigación Científica de Yucatán, Mérida, Yucatán, Mexico

Abstract

A problem of heat conducting and unilateral contact of a shell with a rigid body through the heat-conducting layer is formulated. An approach consists in considering a change of layer thickness in the process of the shell deformation. For modelling thermoelastic state of the shell classical Kirchhoff-Love's model is explored. For modelling heat conductivity of the shell expansion into a polynomial Legendre series in terms of the thickness is used. Contact conditions that take into account possibilities for unilateral mechanical contact and change of heat transfer conditions between shell and rigid body are formulated. Numerical examples of the unilateral thermoelastic contact of the cylindrical shells and rigid body through the heat-conducting layer are considered. Influence of physical and geometrical parameters of the shell and heat conducting layer is investigated.

Keywords: heat-conductivity, cylindrical shell, heat-conducting layer, mechanical contact.

1 Introduction

Contact interaction is the most common way to transfer load from one body to another. In the case of contacting bodies having different temperatures between them heat-contact interactions take place. Therefore, not only the condition of the mechanical contact, but also conditions of the thermal contact have to be considered. Usually, perfect thermal contact is assumed, i.e. it is assumed that the temperature and the thermal flux of the contacting bodies in the contact area are the same [3]. In numerous publications [2, 4–15] it was shown that in many cases these contact conditions are not acceptable because they can not take into

WIT Transactions on Engineering Sciences, Vol 61, © 2008 WIT Press
www.witpress.com, ISSN 1743-3533 (on-line)
doi:10.2495/HT080081

account physical processes related to deformation and heat exchange. In these publications the problem of thermoelastic contact of plates and shells have been considered through a heat-conduction layer with considering change of the layer thickness during the plates and shells deformation. Numerical examples presented there show that in many important cases for science and engineering the result obtained using perfect thermoelastic contact conditions and the conditions with considering change of the layer thickness in the process of deformation are very different. In some cases the difference is not only quantitative but also qualitative. Therefore, it is very important to consider contact conditions which relate deformations and heat transfer in the problems where thin-walled structures may have contact though the heat-conducting layer in the intensive temperature field. Such kinds of problems take place in many important structures, equipment, and devices in chemical, airspace, nuclear industries, etc.

The developed approach have been applied to the plates and shells thermoelastic contact problems in [4–7, 11–15], the laminated composite materials with the possibility of delamination and thermoelastic contact in temperature field in [8, 9], and the pencil-thin nuclear fuel rods modeling in [10].

In this paper some new results related to unilateral thermoelastic contact of the axisymmetrical cylindrical shell through the heat-conducting layer are formulated. The connected equations of thermoelasticity and heat conductivity are created. These equations take into account change of the conditions of heat exchange between the shell and the rigid body during its deformation and possibility of close unilateral mechanical contact. Numerical examples of the heat conductivity of the cylindrical shells through the heat-conducting layer are considered. The thermomechanical effects caused by contact interaction and their influence on the thermomechanics parameters are investigated.

2 Statement of the problem

Let us consider an axisymmetrical cylindrical shell with parameters: r is a radius, $2h$ is a thickness, l is a length, Ω, Ω^+ and Ω^- are middle, external and internal surfaces of the shell. We consider two situations illustrated in the fig.1: a) a rigid punch is placed inside of the shell with gap $h_0(x)$ b) a cylindrical shell placed inside of the cylindrical hole in the rigid body with gap $h_0(x)$.

The heat is transferred from the body surface ψ to the shell and inversely through the heat conducting layer which is not resisted shell deformation. The shell can be subjected to external mechanical load $p(x)$ and temperature $T(x)$. Consequently, the shell is deformed and can come into mechanical contact with the rigid body. As a result is established as unknown before close contact area $\Omega_e = \Omega^{\pm} \cap \psi$ and forces of contact interaction $g(x)$. Thus we have a situation with deformation influence on heat exchange and temperature influence on deformation.

In such formulation we have connected the thermoelastic and heat conducting problem. In this case, equations of thermoelasticity and heat conductivity cannot be solved separately. The possibility for unilateral mechanical contact have to be also taken into account.

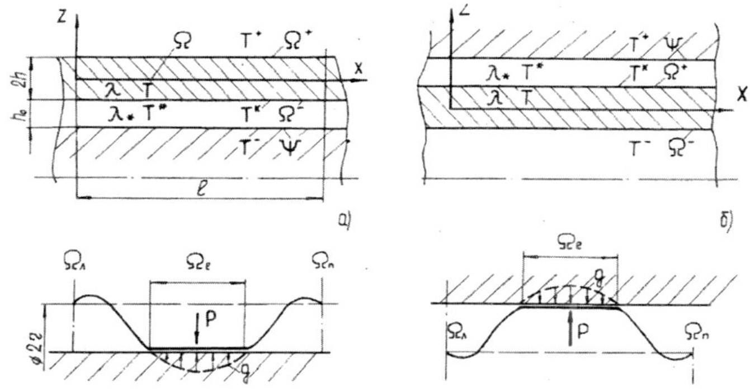

Figure 1.

2.1 Equations of thermoelasticity of the shell

We will use here a classical Kirchhoff-Love's shell theory. According to this theory axial ε_x and circular ε_θ deformations of the shell middle surface are defined by the equations

$$\varepsilon_x = \frac{du + dvr}{dx} = \frac{du}{dx} + \frac{d^2 w}{dx^2} z \, , \; \varepsilon_\theta = \frac{2\pi(r - z + w) - 2\pi(r - z)}{2\pi(r - z)} = \frac{w}{r - z} \approx \frac{w}{r} \quad (1)$$

and corresponding stresses by

$$\sigma_x = \frac{E}{1-\nu^2}\left[\frac{du}{\partial x} + \frac{d^2 w}{\partial x^2} z + \nu \frac{w}{r} - \alpha_t (1+\nu) T\right],$$

$$\sigma_\theta = \frac{E}{1-\nu^2}\left[\nu \frac{du}{\partial x} + \nu \frac{d^2 w}{\partial x^2} z + \frac{w}{r} - \alpha_t (1+\nu) T\right], \quad (2)$$

where u and w are displacements in axial and circular directions, E and ν are modulus of elasticity and Poison ratio, $T = T_0 + T_1 z$ is temperature of the shell. Differential equations of thermo-elasticity have the form

$$\frac{d^4 w}{dx^4} + 4\beta^4 w = \frac{1}{D}(p - g) - a_0 T_0 + a_1 \frac{d^2 T_1}{dx^2}, \quad (3)$$

where $\beta^4 = \frac{3(1-\nu^2)}{4h^2 r^2}$ $a_0 = \frac{3\alpha_t (1-\nu^2)}{h^2 r}$, $a_1 = \frac{\alpha_t (1+\nu)}{h}$

2.2 Equations of heat conductivity of the shell

Following [4] we represent temperature of the shell in the form

$$T = T_0 + \frac{z}{h} T_1 \, , \; T_0 = \frac{1}{2h}\int_{-h}^{h} T dz \, , \quad T_1 = \frac{3}{2h^2}\int_{-h}^{h} T z dz \quad (4)$$

Then differential equations of heat conductivity have the form

$$\frac{d^2T_0}{dx^2}+\frac{1}{2h}\left(Q^+-Q^-\right)-\frac{1}{r}Q_0=0$$

$$\frac{d^2T_1}{dx^2}+\frac{3}{2h}\left(Q^++Q^-\right)-\frac{3}{h}Q_0+\frac{1}{r}Q_1=0 \tag{5}$$

Here T_i and Q_i are coefficients of the polynomial expansion of the temperature and heat flux. They are related by equations

$$Q_0=\frac{1}{2h}\left(T^+-T^-\right),\ Q_1=\frac{3}{2h}\left(T^++T^-\right),\ Q_2=5Q_0-\frac{5}{h}T_1,$$

$$Q^++Q^-=2(Q_0+Q_2),\quad Q^+-Q^-=2Q_1 \tag{6}$$

where T^+,T^-,Q^+ and Q^- are their values on the surfaces Ω^+ and Ω^- respectively.

2.3 Boundary and contact conditions

For finite length shell we consider the following mechanical boundary conditions.

$$\left.\frac{d^3w}{dx^3}\right|_{\Omega^L,\Omega^P}=0,\ \left.\frac{d^2w}{dx^2}\right|_{\Omega^L,\Omega^P}=0 \text{ - free end; } w|_{\Omega^L,\Omega^P}=0,\ \left.\frac{dw}{dx}\right|_{\Omega^L,\Omega^P}=0 \text{ - fixed end;}$$

$$w|_{\Omega^L,\Omega^P}=0,\ \left.\frac{d^2w}{dx^2}\right|_{\Omega^L,\Omega^P}=0 \text{ - simply supported end.} \tag{7}$$

For infinite length shell we consider the following conditions at infinity

$$w\to 0,\ \frac{dw}{dx}\to 0,\ \frac{d^2w}{dx^2}\to 0,\ \frac{d^3w}{dx^3}\to 0 \text{ for } x\to\infty \tag{8}$$

Boundary conditions for the equations of heat conductivity are

$$T|_{\Omega^L}=T^L \text{ - if temperature is prescribed;}$$

$$\lambda\left.\frac{dT}{dx}\right|_{\Omega^L}=q^L \text{ - if heat flux is prescribed;} \tag{9}$$

$$\left.\frac{\partial T}{\partial x}+\frac{\alpha}{\lambda}\left(T-T^L\right)\right|_{\Omega^L}=0 \text{ - for convective heat transfer}$$

where λ is a coefficient of heat conductivity, α is a coefficient heat exchange though a surface.

Mechanical and thermal contact conditions are presented in the form [4–7]

$$w<h_0\to g=0\ ,\ \forall x\in\Omega^+\setminus\Omega_e\quad ,\ w=h_0\to g>0\ ,\ \forall x\in\Omega_e$$

$$T=T_k\ ,\ T_k=\frac{\lambda(h_0-w)T^++\lambda_*hT^-}{\lambda(h_0-w)+\lambda_*h} \tag{10}$$

It is important to mention that thermal contact conditions include shell deflection w nonlinearly and therefore equation (10) relates the equations of shell thermoelasticity (3) and heat conductivity (5). As a result we have connected nonlinear problems of thermoelasticity and heat conductivity with unilateral contact conditions.

2.4 Transformation to the integral equation

In [4, 6, 7, 13–15] it was shown that the differential equations of the shell thermoelasticity (3) and heat conductivity (5) can be transformed into the integral equations of Hammerstein's type

$$\int_l G_0(x,y)F_0(y)dy = T^0 \;,\; \int_l G_1(x,y)F_1(y)dy = T^1$$

$$\int_l W(x,y)\left\{\frac{1}{D}\left[p(y)-g(y)\right]-\beta_0 F_3(y)\right\}dy = w\,, \tag{12}$$

where

$$F_0 = 0.5\varepsilon_0\left(T^- + T_k\right)+\frac{1}{2hr}\left(T_k - T^-\right),\; F_1 = 0.5\varepsilon_1\left(T_k - T^-\right)+\frac{3}{2hr}\left(T_k + T^-\right)-\frac{3}{hr}T^1,$$

$$F_3 = \beta_1\left(F_1 + \varepsilon_1^2 T^1\right)-\beta_0 T^0$$

The kernels in these integral equations are fundamental solutions for corresponding differential operators. They have the form

$$G_0(x,y)=\exp\left(-\varepsilon_0|x-y|\right)/2\varepsilon_0 \;,\; G_1(x,y)=\exp\left(-\varepsilon_1|x-y|\right)/2\varepsilon_1 \tag{13}$$

$$W(x,y)=\frac{1}{8\beta^3 D}\exp\left(-\beta|x-y|\right)\left[\cos\left(\beta|x-y|\right)+\sin\left(\beta|x-y|\right)\right]$$

where

$$\varepsilon_0 = \frac{3}{h^2},\; \varepsilon_1 = \frac{15}{h^2},\; \beta_0 = \frac{3(1-\nu)\alpha_\tau}{h^2 r},\; \beta_1 = \frac{(1+\nu)\alpha_\tau}{h},\; D=\frac{2Eh^3}{3\left(1-\nu^2\right)}$$

2.5 Algorithm for the problem solution

Algorithms for the problem's solution consists in an iterative process of the nonlinear integral equations of Hammerstein's type solution and in the case if unilateral contact taking place an additional iterative algorithm is used. The algorithm has been elaborated in [4, 6]. In the problems under consideration the algorithm is convergent and convergence is fast enough.

2.6 Stress calculation

Stresses in the axisymmetrical cylindrical shell are calculated by formulas (2), which for convenience may be presented in the form

$$\sigma_x = \frac{E}{1-\nu^2}\left[\frac{d^2 w}{dx^2}z-(1+\nu)\alpha_t T^1\frac{z}{2h}\right], \tag{14}$$

$$\sigma_\theta = \frac{Ew}{r}-\alpha_t T^0 E+\frac{E}{1-\nu^2}\left[\nu\frac{d^2 w}{dx^2}z-(1+\nu)\alpha_t T^1\frac{z}{2h}\right]$$

Deriving from (12) expression for the second derivative of the displacements and substituting it into (14) we obtain the following integral representations for stress

$$\sigma_x(x) = b_0 \int_0^l T^0(\xi) \frac{d^2G(\xi,x)}{dx^2} d\xi + \tag{15}$$

$$b_1 \int_0^l G(\xi,x) T^1(\xi) d\xi - b_2 T^1(x) + b_3 \int_0^l [p(\xi) - q(\xi)] \frac{d^2G(\xi,x)}{dx^2} d\xi$$

$$\sigma_\theta(x) = \nu\sigma_x(x) + \frac{Ew(x)}{r} - \alpha_t E\left[T^0(x) + T^1(x)z\right]$$

where

$$b_0 = \frac{3E\alpha_t zr^5}{(1+\nu)h^2 h_0}, \quad b_1 = \frac{3(1+\nu)E\alpha_\tau r^6 z}{4h^3 h^2{}_0}, \quad b_2 = \frac{E\alpha_\tau r^4 z}{(1-\nu)hh_0},$$

$$b_3 = \frac{3r^6 z}{2h^3 h_0}, \beta_0 = \frac{3\alpha_t Erz}{h^2}, \quad \beta_1 = \frac{3(1-\nu)\alpha_t Er^2 z}{h^2}, \quad \beta_3 = \frac{r^2 z}{8h^3}.$$

Mizes stresses are calculated by the equations

$$\sigma_i^- = \sqrt{(\sigma_x^-)^2 + (\sigma_x^-)(\sigma_0^-) + (\sigma_0^-)^2}, \tag{16}$$

$$\sigma_i^+ = \frac{1}{\sqrt{2}}\sqrt{(\sigma_x^+ - q)^2 + (\sigma_x^+ - \sigma_x^+)^2 + (\sigma_0^+ + q)^2}$$

Using equations (15) and (16) stresses in the shell can be easy calculated.

3 Investigation of the thermo-mechanical state of the shell

We will consider here thermo-mechanical state of the shell with considering influence of the shell deformation on the heat transfer between it and a rigid body. Calculation have been done for the data: material properties: $E = 2.5 \cdot 10^5$ MPa, $\nu = 0.25$ $\alpha_\tau = 2.5 \cdot 10^{-5}\, 1/{}^oC$, $\lambda_* = 1V/m^oC$ and geometrical parameters $r = 0.5m$, h=0.01m, $l_b = r$,

Example 1. We consider an axisymmentrical cylindrical shell of infinite length placed into the rigid stirrup with a homogeneous initial gap as is shown in fig.2. Temperature on the stirrup surface is not homogeneous and equal to

$$T^+(x) = T_m + T_b \sin \pi x / l_b, T_b = 600^o C \; T_m = 100^o C. \tag{17}$$

On the shell surface act homogeneous load $p(x) = 10MPa$, temperature on the internal surface of shell is $T^- = 0^o\ C$, initial gap is equal to $h_0 = 0.5h$ and $\lambda_1 = 2\lambda_*$.

In fig.3 are presented: Mizes stresses on external σ^+ and internal σ^- surfaces of the shell, force of contact interaction g, normalized bending $W = w / h_0$ and temperature on contact surface T_k. The dashed lines correspond to a solution for perfect thermal contact without counting influence of the shell deformation on the heat exchange and the solid lines correspond to the solution presented here.

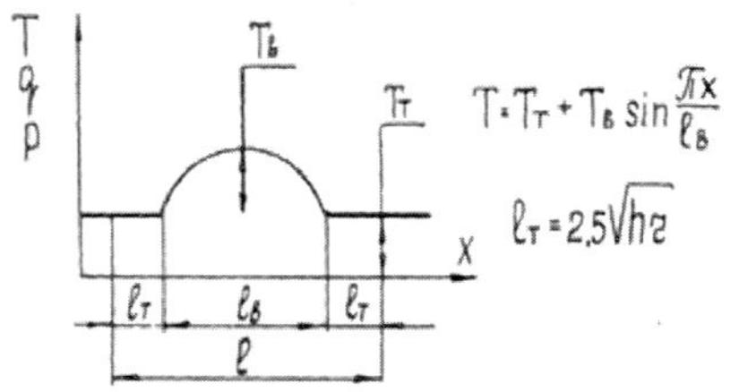

Figure 2.

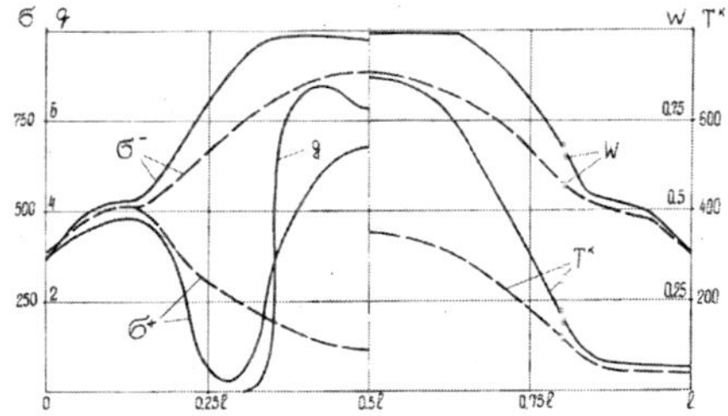

Figure 3.

Mizes stresses on external σ^+ and internal σ^- surfaces of the shell, force of contact interaction g, for the same data and initial gap $h_0 = 0.1h$ are presented in fig 4 and 5 external and internal punch respectively.

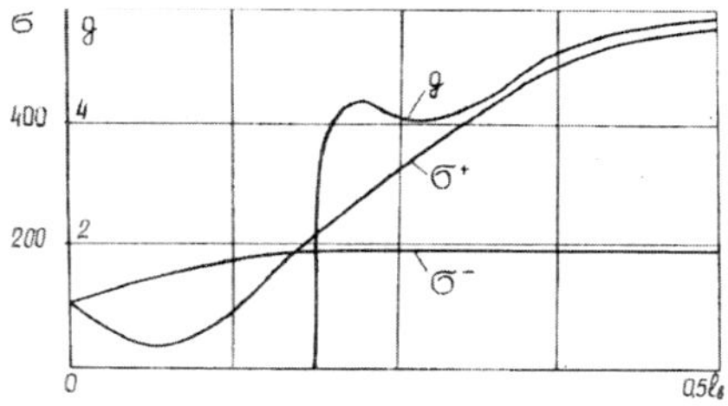

Figure 4.

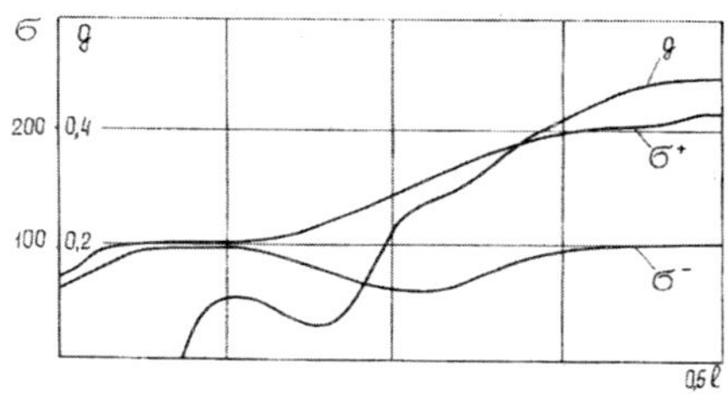

Figure 5.

Analysis of these data shows that no homogeneous temperature distribution cause significant shell deformations and close mechanical contact with a rigid stirrup. As a result significant stress occurs in the shell. Calculations assuming perfect thermal contact lead to significant inaccuracy, which is not only quantitative but also qualitative.

Example 2. Here we consider an axisymmentrical cylindrical shell of infinite length placed into the rigid stirrup with no homogeneous initial gap, as shown in the Fig.6.

The gap is given by the function

$$h_*(x) = h_o + h_b \sin \pi x / l_b \,, h_0 = 0.5h \,, h_b = h_0 / 2 \,, \; l_b = r \,.$$

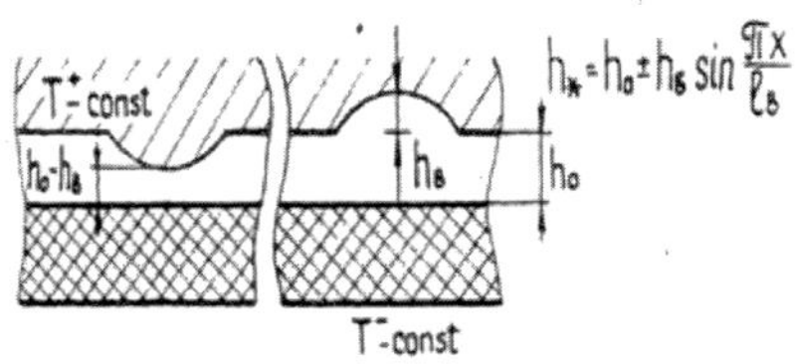

Figure 6.

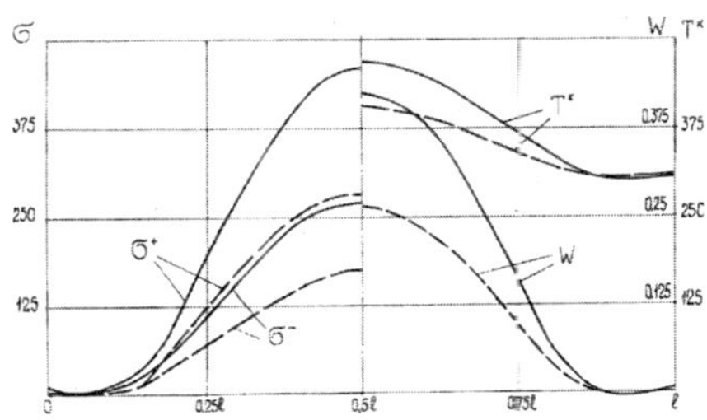

Figure 7.

In fig.7 are presented: Mizes stresses on external σ^+ and internal σ^- surfaces of the shell, normalized bending $W = w / h_0$ and temperature on contact surface T_k. The dashed lines correspond to solution for perfect thermal contact without counting influence of the shell deformation on the heat exchange and the solid lines correspond to the solution presented here. Analysis of these data shows, that no homogeneous initial gap cause significant shell deformations. As a result, in the shell, no significant homogeneous stress and temperature distribution occurs. Calculations assuming perfect thermal contact lead to significant inaccuracy. Some values of thermomechanical state differ.

Example 3. We also investigate influence of different parameters on shell deformation and heat exchange. In fig. 8 are presented deformations and temperature on the contact surface for fixed-end shell in homogeneous temperature field $T^- = 150^o\ C$, $T^+ = 0^o\ C$, for $\lambda_1 = 4\lambda_*$, $h_0 = 0.25h$.

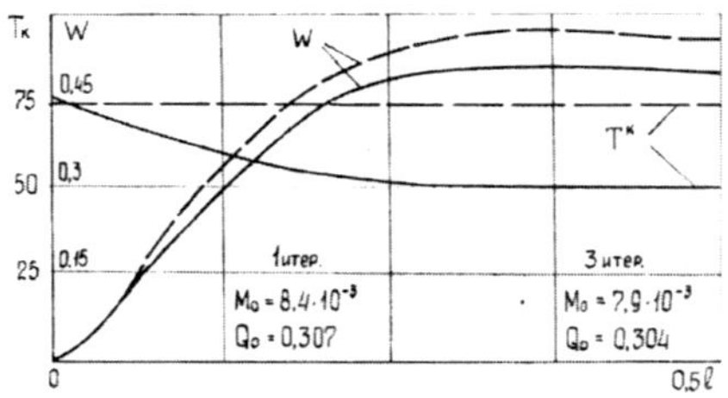

Figure 8.

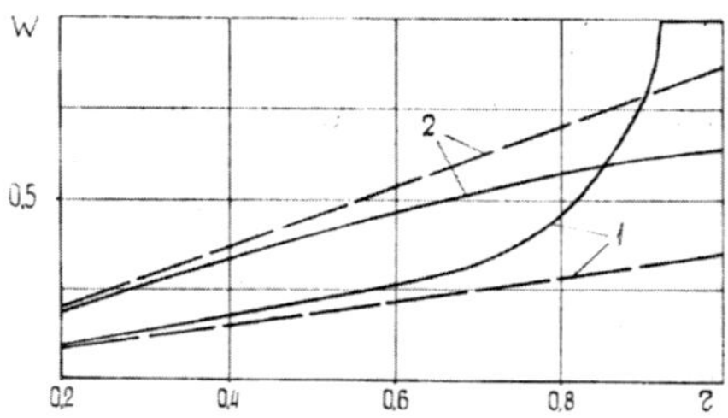

Figure 9.

It is interesting to point out that in the case when $T^- > T^+$ accounting influence of the shell deformation on the heat exchange decreases its deformation, stress and temperature distributions.

In fig. 9 is presented dependence of the shell deflection on its radius. Calculations have been done for the following data: $\lambda_1 = 5\lambda_*$ $h_0 = 0.5h$, curve 1 for the external punch and $T_b = 600^o C$, curve 2 for the internal punch and $T_b = 200^o C$. It is important to mention that for external punch when $r / h > 50$ dependence becomes nonlinear and the shell is attracted to the punch, solid curve 1.

All the above calculations have been done for the case when on the surfaces Ω and ψ is prescribed temperature. In fig. 10 and fig. 11 dependences $w = f(\lambda_1 / \lambda_*)$ and $w = f(h_0 / h)$ for the external punch are presented. Curves 1 correspond to the case when on the surfaces ψ temperature and on the surface Ω heat flux are set. Curves 2 correspond to the case when on the surfaces Ω temperature and ψ on the surface heat flux are set. In both cases temperature is distributed in accordance with equation (17), heat transfer rate is equal to $q = -2 \cdot 10^4 [V / m^2 K]$ and $q = 1.4 \cdot 10^4 [V / m^2 K]$ for the first and second cases respectively. In the first case shell deflection is minimal for $\lambda_1 / \lambda_* = 1$ and increases when h_0 / h decrease. In the second case shell deflection is maximal for $\lambda_1 / \lambda_* = 1$ and does not depend much on h_0 / h. In both cases shell deflection does not change much for $\lambda_1 / \lambda_* > 10$.

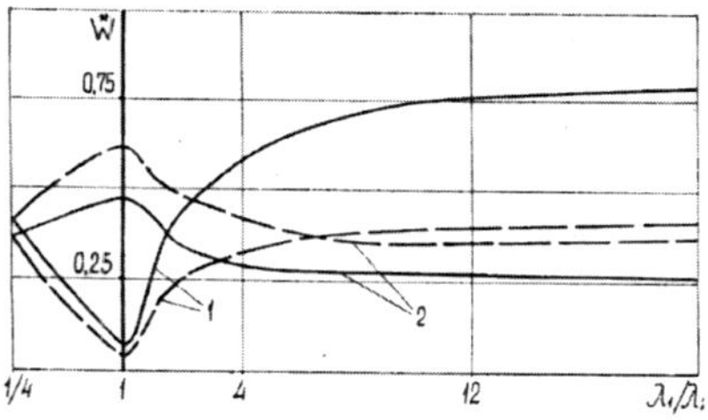

Figure 10.

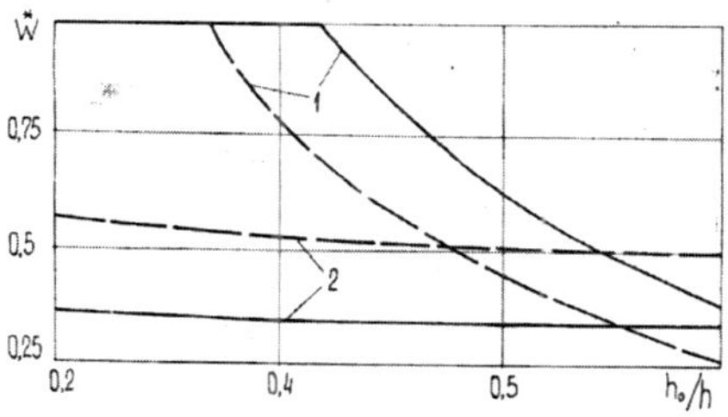

Figure 11.

4 Conclusions

The results presented here show that in many situations thin-walled structures subjected to high temperature perfect thermal contact conditions are not acceptable because they cannot take into account physical processes related to deformation and heat exchange. Such kinds of problems may occur in different field of science and engineering, for example in nuclear, aerospace, chemical industries, etc. In such situations the approach developed here and in our previous publications have to be used.

References

[1] Boley B.A. and Weiner J.H. *Theory of thermal stress*, Wiley, New York, 1960.

[2] Kantor B. Ya., Zozulya V.V. Connected problem on contact plate with rigid body though the heat-conducting layer, *Doclady Akademii Nauk Ukr. SSR*, **4**, pp. 31–33, 1988. (in Russian)

[3] Postrigach Ya. S, Shvets P. N. Yu. M., *Thermoelasticity of thin shells*, Kiev, Naukova dumka, 1978. (in Russian)

[4] Zozulya V.V. Contact cylindrical shell with a rigid body though the heat-conducting layer, *Doclady Akademii Nauk Ukr. SSR*, 1989, **10**, pp.48–51. (in Russian)

[5] Zozulya V.V. The combined problem of thermo- elastic contact between two plates though a heat conducting layer, *Journal Applied Mathematics and Mechanics*, **53**(5), pp.722–727, 1989.

[6] Zozulya V.V. Bending of a plate in temperature field under restrictions, *Izvestiya vuzov. Engineering*, **1**, pp. 24–27, 1990. (in Russian)

[7] Zozulya V.V. Contact cylindrical shell with a rigid body though the heat-conducting layer in transitional temperature field, *Mechanics of Solids*, Vol. 2, pp.160–165, 1991.

[8] Zozulya V. V. Nonperfect contact of laminated shells with considering debonding between laminas in temperature field, *Theoretical and Applied Mechanics,* Vol. 42, pp.92–97, 2006.

[9] Zozulya V.V. Laminated shells with debonding between laminas in temperature field, *International Applied Mechanics,* Vol. 42(7), pp. 135–141, 2006.

[10] Zozulya V.V. Mathematical Modelling of Pencil-Thin Nuclear Fuel Rods, in *Proc. SMIRT19 Conf. Structural Mechanics in Reactor Technology,* A. Gupta Ed., Toronto, 2007, pp. C04–C12.

[11] Zozulya V.V. Contact of a shell and rigid body though the heat-conducting layer temperature field, International *Journal of Mathematical Models and Methods in Applied Sciences,* **2**(1), pp.138–145, 2007.

[12] Zozulya V.V. Contact of the thin-walled structures and rigid body though the heat-conducting layer, in *Proc. Theoretical and Experimental aspects of heat and mass transfer,* Acapulco, Mexico, pp. 145–150, 2008.

[13] Zozulya V.V., Aguilar M., Thermo-elastic contact and heat transfer between plates through the heat-conducting layer, in *Heat transfer 2000*, B. Sunden and C.A. Brebbia Eds. Computational Mechanics Publications, Southampton, UK and Boston, USA, pp. 123–132, 2000.

[14] Zozulya V.V., Borodenko Yu. N. Thermoelastic condition of cylindrical shell, which interaction with a rigid body though the heat-conducting layer, *Izvestiay vuzov. Engineering*, **8**, pp. 47–52, 1990. (in Russian)

[15] Zozulya V.V., Borodenko Yu. N. Connecting problem on contact of cylindrical shells with a rigid body in temperature though the heat-conducting layer, *Doclady Akademii Nauk Ukr. SSR*, **4**, pp.35–41, 1992. (in Russian).

Section 4
Heat recovery

An experimental feasibility study of using diesel exhaust for space heating in Alaskan Villages

P. Raghupatruni, C.-S. Lin, D. Witmer, E. Bargar, J. Schmid, T. Johnson & V. Avadhanula
Department of Mechanical Engineering, University of Alaska Fairbanks, USA

Abstract

In rural Alaska, there are nearly 180 villages consuming about 374,000 MWh of electrical energy annually from individual diesel generators. A similar amount of fuel energy is dissipated into the atmosphere from diesel exhaust. Due to the isolation and small populations of the villages, the costs of fuels, components, and maintenance are extremely high and one of the policies in management is to minimize the possibilities of power outrage and the frequency of maintenance. Also due to the potential of corrosion and soot accumulation in exhaust heat recovery systems, exhaust heat recovery has long been avoided. Recently, the surge in fuel prices, the new regulation requiring ultra low sulfur diesel, and the desire to reduce green house gas has led to exhaust heat recovery being reconsidered for improving the energy efficiency of rural Alaskan diesel power plants.

This report discusses the selection of different exhaust heat recovery applications to the villages, the design of an experimental system that simulates different space heating methods used in villages, system fabrication, instrumentation, system performance, feasibility and economic analysis. Performance presented includes the effect of the heat recovery system on engine performance, heat recovery effectiveness, measured soot accumulation and the effect of fouling on heat recovery effectiveness, and the estimated maintenance requirement.

Keywords: exhaust heat recovery, experimental simulation, space heating, soot accumulation, feasibility, payback time.

www.witpress.com, ISSN 1743-3533 (on-line)
doi:10.2495/HT080091

1 Introduction

Rural Alaskan villages consume about 374,000 MWh of electric energy from individual diesel generators [1]. The single largest amount of unused heat from engines is exhaust heat, which contains about 30% of the fuel energy. Work which studies the selection of the most appropriate engine heat recovery method is needed as part of the village economic development. Several heat recovery applications have the potential to be economic and feasible for Alaskan villages, such as desalination [2], ice making [3], direct thermal electric conversion [4], heat to power conversion [5–7], and space and water heating [8]. Desalination did not appear to be locally useful due to the mineral contents of ground water for most of the Alaskan villages [9], which were well within acceptable ranges. Direct thermal to electricity conversion was not selected due to its low conversion efficiency. Ice making has been demonstrated in one of the villages, but unless there is a large commercial user for ice, the cost was not justified.

Both heat to power conversion and space and water heating were expected to have reasonable recovery efficiencies. Heat to power conversion can easily transform the heat into a convenient form of energy. Based on some current reports [5–7, 10], the expected improvement in fuel efficiency of different heat to power conversion applications were about 10%. However, Most of the development for medium size diesel engines, which were the majority of the engines used in Alaskan villages, is still in the research and prototyping stages. Based on the efficiencies of the readily available components, an exhaust heat recovery system for space and water heating could be easily fabricated to deliver 50% or more of the exhaust heat energy (i.e. 15% or more of fuel energy) [11].

This project studies the feasibility and economic effect of applying exhaust heat recovery for space heating to Alaskan village diesel generators. Heat recovery for thermal to power would be the subject of the next study. Space heating, which included public space heating using baseboard and floor radiant systems, were also proposed by the Alaska energy authority as candidates for future rural Alaska cogeneration development [8]. The following sections describe details of the diesel generator and the heat recovery system used for this study along with system instrumentation, experimental results, and the final conclusion.

2 Test bed

This section discusses the test bed and the details of the heat recovery system.

2.1 The diesel generator system

This experiment was conducted using a DD50 diesel-generator set, one of the typical diesel generators used in villages. The generator was connected to a 250 kW external resistive/reactive load bank for engine load control and simulation. Load bank was interfaced with LABVIEW IV system for load control signals. The diesel generator set and data acquisition system were placed inside an ISO container while the load bank sitting outside of the ISO container. Details can be found in [12]. Property of the diesel engine is listed in Table 1.

2.2 Exhaust heat recovery system

The exhaust heat recovery system was designed according to the following requirements:

Apply existing technology.

Optimize overall efficiency for varying engine load (In general, the load of a village generator may range from 25% to 100% of rated load).

Minimize corrosion to exhaust system.

Meet back pressure requirement (13.8 kPa or 2 psi).

Meet dimension constraints (very limited space).

Be able to emulate different space heating applications of Alaskan villages.

Be able to measure system and component performance.

Be easy to maintain and do not increase maintenance frequency.

Work under a wide range of ambient temperatures (-40°C to 33°C).

Be reliable.

Table 1: Diesel engine properties.

Number of cylinder: 4	Number of Stroke: 4
Bore x Stroke: 130mm x 160mm	Fuel Injection: Electronic unit injector
Aspiration: Turbocharged	Displacement: 8.5 L
Rated Power: 125 kW	Rated Speed: 1200 rpm
Exhaust Gas Recirculation: None	Exhaust After treatment: None

Based on the requirements, an experimental heat recovery system was designed and installed. The system had three major sections: the heat source section (Figure 1A), the heat sink or load section (Figure 1B), and the pipe and control section (Figure 1C). The source section had a heat exchanger and auxiliary components to capture heat from engine exhaust. The heat sink section had a unit heater and auxiliary components to simulate the village load and was designed to work in such a way that the source or load could be controlled at specific temperatures at different engine loads. The pipe and control section was used to transport working fluid between the heat source and the heat sink and able to control the flow rate and pressure distributions of the system.

Based on the computation results [13, 14] and the availability of commercial products for high temperature exhaust, a shell and tube type of heat exchanger was selected. The exhaust gas passed through the shell side and the liquid coolant through the core side. For maintenance purposes, the core was removable for cleaning when necessary. In order to have the heat exchanger be able to absorb the most amount of heat, the designed size of the heat exchanger was based on the exhaust condition at full engine load. For partial engine load conditions, the heat recovery system was designed with parameters adjustable to make the coolant inlet or outlet temperature match the requirements of different heating applications. The heat exchanger coolant inlet side temperatures used for design (Table 2) were selected based on the feed water temperature of 65°C typical of boilers used for baseboard heating in rural Alaska [8]. So a temperature of 77°C would be adequate for pre-heating the boiler feed water. The inlet gas temperature used for design was the full load exhaust temperature and the outlet gas temperature was based on the acid condensation formation temperatures [14, 15] of the exhaust and also based on the trade off between the cost and the amount of heat to be recovered. Some of the details of the selected heat exchanger are shown in Table 3.

Figure 1: The heat recovery system. (a): Heat exchanger section (top of container). (b): Unit hater section (outside of container). (c-1): Pipe and control elements (south half). (c-2): Pipe and control elements (north half).

The heat dissipating capacity of the selected unit heater was above the expected maximum amount of heat absorption rate of the heat exchanger with a safety margin. A bypass loop and a 3 way temperature control valve were selected for the control of the outlet temperature of the load section by appropriately distributing the total flow between the bypass loop and the unit heater.

The pipe and control section included the pipes between the heat exchanger section and the load section, a pump and other miscellaneous components (i.e. ball valves, expansion tank, circuit setters, snubbers, dielectric unions, fittings, insulation material, etc.) were designed to control the system operation and to meet the needs for maintenance, instrumentation, corrosion prevention, etc. The pump was selected based on the design flow rate and calculated total pressure drop [16] of the system with the effects of all the expected system components.

3 Instrumentation and test procedure

3.1 Instrumentation

The purpose of instrumentation was for system operation monitoring and system data collection. A National Instrument data acquisition system (DAQ) was used to document the experimental readings. Temperature data were used to estimate the system and components efficiencies, possibility of condensation formation in exhaust pipe and heat exchanger, and effect of fouling on system performance. Pressure data were used to monitor and evaluate the condition of the flow path and the functions of the components. The locations of the temperature sensors and the coolant flow meter along the pipe system are shown in Figure 2. There were also 5 thermal couples installed at the heat exchanger exhaust outlet to measure the instantaneous average temperatures of exhaust and 10 thermal couples installed on the outside walls of the heat exchanger core tubes to monitor the temperature drop across the heat exchanger. The exhaust temperature at the inlet of the heat exchanger was obtained from a sensor came with the original engine generator set.

Table 2: Inlet/outlet temperatures of heat exchanger.

	Inlet temperature	**Outlet temperature**
Exhaust	540 ^{0}C	177 ^{0}C
Coolant	77 ^{0}C	87 ^{0}C

Table 3: Heat exchanger parameters.

Shell side	Gas
Tube side	Liquid
Gas pressure drop	0.31 PSI
Maintenance	Removable core
Heat transfer area	87 ft^2
Size	51"x28"x28"
Weight	725 lbs
Material of construction	Tube- SS fin tube Shell- SS inner wall
Insulation	Integrated insulation

In addition to the measurements mentioned above, engine inlet air flow rate and property (i.e. temperature, pressure, humidity) and engine fuel consumption rate were also measured and used for exhaust flow rate calculations. Details can be found in [12, 17]. Sensors for mass flow rates of fuel consumption and coolant flow were calibrated gravimetrically.

3.2 Test procedure

The heat recovery system was operated and monitored for nearly 350 hours after installation. In order to observe the worst case scenario, the fuel used for testing was the conventional diesel fuel and the engine was not equipped with any of the after-treatment devices. The system was operated under generator rated load conditions (125 KW) for most of the time. The system was also operated under different engine loads (25%, 50%, and 75%) for performance testing. For each load, the heat recovery system was tested for three different types of space heating applications defined by the coolant outlet temperatures of the heat exchanger [8].

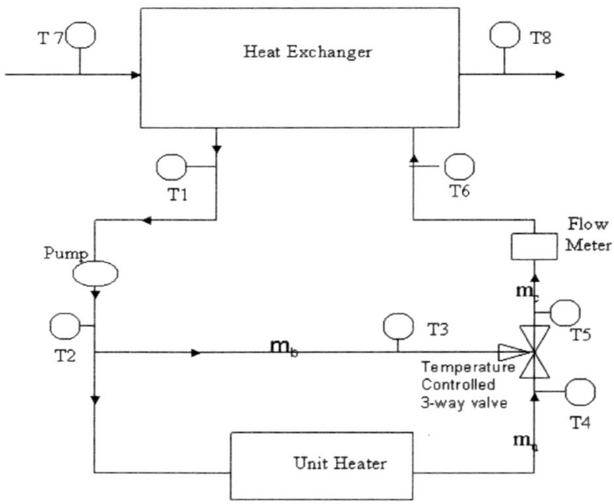

Figure 2: Locations of temperature sensors and flow meter.

During the first 150 hours of test conducted in summer 2006, water was used as coolant to investigate the defects in installation and calibrations and the effect of adding the heat recovery system on general engine performance. At the 150th hour, the coolant was changed from water to 40% propylene and 60% water mix to avoid freezing and the flow meter was recalibrated. The next 100 hours test, conducted in fall 2006 was to investigate the measurement reliability and performance consistency. After the 250th hour, one of the thermocouples was recalibrated and the time constant of the temperature control valve was adjusted to limit the fluctuation in the temperature readings.

Data collected during the last 100 hours test conducted in the winter of early 2007 was used for most of the analysis because the data were considered more accurate. In this report, the analysis results were mostly based on the data collected between the 250th hour and 300th hour. During this period, the engine was operated under full load and data was used to investigate the general system performance, feasibility, and payback time. Between the 300th hour and 350th hour, the engine was operated under 4 different engine loads (i.e. 25%, 50%, 75%, and 100% of rated load) and data was used to investigate the effect of the load on the performance of the heat recovery system.

4 Results and discussion

This section discusses the measured data and analysis results.

4.1 System performance

Based on the first 100 hours test data, no engine performance was found significantly altered. Some of the data is listed in Table 4. Performance data of the exhaust heat recovery system obtained during the period between the 250th hour and 300th hour is given in Table 5.

Table 4: Engine performance data before installation of the heat recovery system and at the 50th hour and 100th hour.

Engine hours	Exhaust T (Degree C)	Turbocharger T (Degree C)	Fuel consumption (L/hr.)
Before the installation	542	122	34
50th hour	540	142	34
100th hour	533	140	34

Table 5: Performance data of the exhaust heat recovery system during the period between 250-hr and 300-hr.

Measured Data	10Hr	20Hr	30Hr	40Hr	50Hr
Exhaust Flow (Kg/s)	0.25	0.24	0.24	0.24	0.25
Exchanger exh inlet T (C)	511.5	503.0	487.0	495.9	517.1
Exchanger exh outlet T C)	215.6	220.6	207.9	214.5	226.6
Coolant flow rate (Kg/s)	1.45	1.48	1.45	1.45	1.43
Exchanger coolant inlet T (C)	76.89	76.53	77.07	77.03	75.82
Exchanger coolant outlet T (C)	88.45	87.05	87.74	87.53	87.34
Before bypass (load) T (C)	88.02	86.79	87.36	87.00	86.83
After 3way valve (load) T (C)	76.91	76.94	77.08	76.75	75.52
Flow rate in bypass (Kg/s)	1.00	1.13	1.02	1.00	0.83
Flow across unit heater (Kg/s)	0.45	0.35	0.43	0.45	0.60
Outside ambient T (C)	-23.0	-30.0	-18.0	-15.0	-7.0

The performance data showed that the ambient temperature has a noticeable effect on the flow distribution of the system between the bypass and the unit heater (Figure 3) as expected. It also shows that the heat dissipated from the unit heater was higher for higher ambient temperatures. This can be explained as, at higher ambient temperatures, the pipeline dissipates less heat that needed to be compensated by increasing heat loss through the unit heater to maintain the inlet temperature of the heat exchanger. The energy balance was checked between the heat absorbed from the exhaust side and heat dissipated from the load side. Figure 4 shows that the heat absorbed by coolant equals the sum of heat dissipated with an error of less than 3%.

Figure 4 also shows the heat release rate from the exhaust and the heat absorption rate of the coolant. The heat release rate of the exhaust across the heat exchanger was based on the enthalpy change of exhaust components after combustion. The heat absorption rate was evaluated using the coolant mass flow rate, specific heat and the temperature difference between the heat exchanger inlet and outlet flow. The amount of heat absorbed by the coolant followed the same trend as the heat released from exhaust across the heat exchanger.

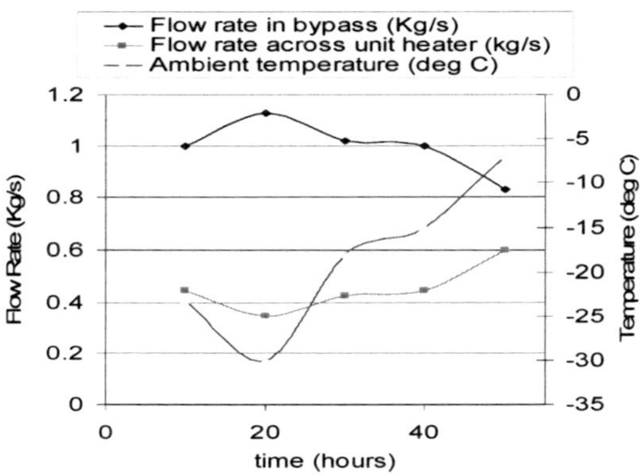

Figure 3: Flow distributions between the bypass and the unit heater.

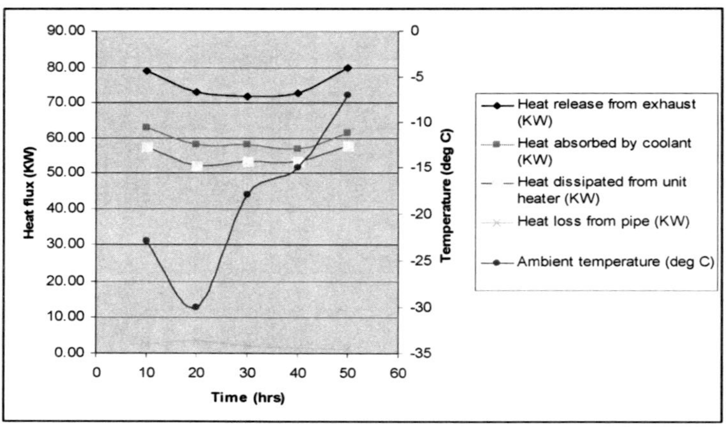

Figure 4: Heat absorption and dissipation with respect to the ambient temperature.

4.2 Soot accumulation and corrosion

Both soot accumulation and corrosion were considered critical parameters, as they were known to have negative effect on the performance and maintenance of heat recovery systems. This section presents the results obtained from the observation of soot accumulation and corrosion inside the heat exchanger after 350 hours of operation of the exhaust heat recovery system.

Over a span of 350 hours of run time, the total soot produced by the exhaust gas was expected to be 3000 to 6000 grams as much of the PM passes through the heat exchanger. After the experiments, the heat exchanger was dismantled and cleaned and about 150 grams of soot was found to be accumulated. Figure 6 shows the soot accumulated on an arbitrarily selected finned tube. This result matches a general believe that soot accumulated on finned tubes may approach to its asymptote within a relatively short period of time [18, 19]. The heat exchanger overall heat transfer coefficients estimated using the experimental data are shown in Figure 7. The values of the overall heat transfer coefficients, which may be affected by the transient values of exhaust gas flow rate and ambient temperatures and fouling resistance, show no monotonic decreasing trend. The trend doesn't conflict with the general believe. Also based on experimental data, estimated average fouling resistance and gas side heat transfer resistance were 0.0105 and 0.0226 m^2-$^oK/W$ respectively. With this small amount of soot accumulated and the simple process needed for cleaning, the heat exchanger needs maintenance not more than twice a year, which is considered no impact to the maintenance frequency of the diesel generator set.

When the heat exchanger was dismantled, the corrosion effect of exhaust gas on the heat exchanger was also investigated by examining the surface for the existence of corrosion spots on both tube and shell sides. No trace of any such corrosion was observed. The reason seems to be that the exhaust temperature was always kept above the acid dew point. This resulted in an absence of condensate in the heat exchanger throughout the run time. The material, SS 316L, used to construct the inside of the heat exchanger may also have the contribution.

Figure 5: Soot accumulations on the fins.

4.3 Economic analysis

The economic analysis was done based on the present heat recovery system and diesel generator used for this experiment and an assumption of 100% use of the recovered heat. The heat recovery rate was taken to be 60 kW at rated load with 8 hours per day. Other parameters used in this calculation include:

Initial cost of the recovery system = $30,000.
Installation cost = $ 5,000 ($75/hr x 8 hrs /day x 10 days).
Airfare, lodging, meals = $1,950 ($600/round trip + $90/day x 15 days).
Total capital cost = $37,950.
Heat recovery per hour = 204,728 BTU.
Heating value for conventional fuel = 130,000 BTU/gal.

The heat recovery system maintenance cost was based on one day of labor ($75/hr) and a round trip flight ticket ($600), which comes to $1,200 for 6 months of engine operation. The maintenance costs also include additional money ($300) every year for supplies.

Pay back time was calculated (Figure 8) with respect to different fuel prices based on two different interest rates and an engine operation of 8 hours per day. The payback time for a 100% use of recovered heat would be less than 3 years based on the fuel price of $3.50/gallon, 10% interest rate, and an engine operation of 8 hours per day.

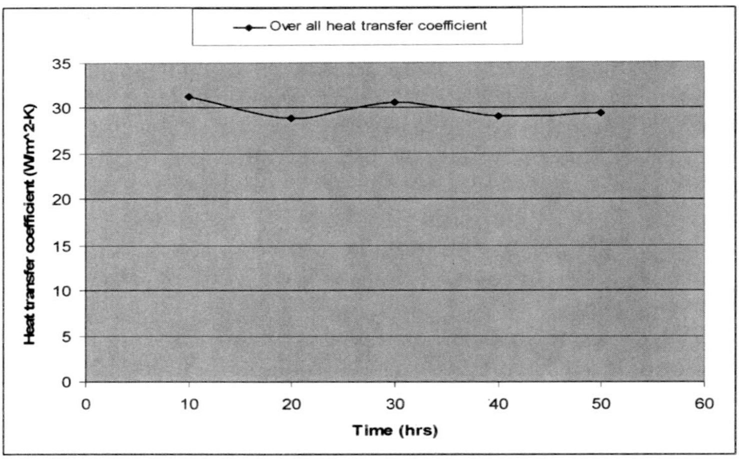

Figure 6: Overall heat transfer coefficient versus time.

5 Conclusions

An experimental work was conducted to check the feasibility and economic effect of applying exhaust heat recovery to a diesel engine generator system. The collected data and analysis results lead to the following conclusions:

1. The performance of our exhaust heat exchanger was reliable and consistent.

2. For the 125 kW diesel generator used in this experiment, the rate of heat recovered from the exhaust was about 60 kW.
3. No effects were observed on the engine performance and maintenance frequency due to the heat recovery system.
4. According to the soot accumulation data obtained from this experiment, the estimated time for heat exchanger maintenance is less than two days per year.
5. Corrosion was not observed to be a problem in the laboratory test of 350 hours.
6. Based on experimental data obtained from this experiment, the estimated payback time for a 100% use of recovered heat would be less than 3 years for a fuel price of $3.5 per gallon and an engine operation of 8 hours per day. For 80% use of the recovered heat, the payback time would be 4 years.
7. Operation cost is largely case dependent. Influential parameters would include diesel fuel cost, the application of the recovered heat, location of the power plant, etc.
8. Performance and economic outcomes will be different from one case to another. For example the existing infrastructure of the community water heating system may largely affect the capital cost due to the high cost of the outdoor arctic tubing and installation requirement. However, analysis is recommended before the installation of an exhaust heat recovery system to a village generator set.

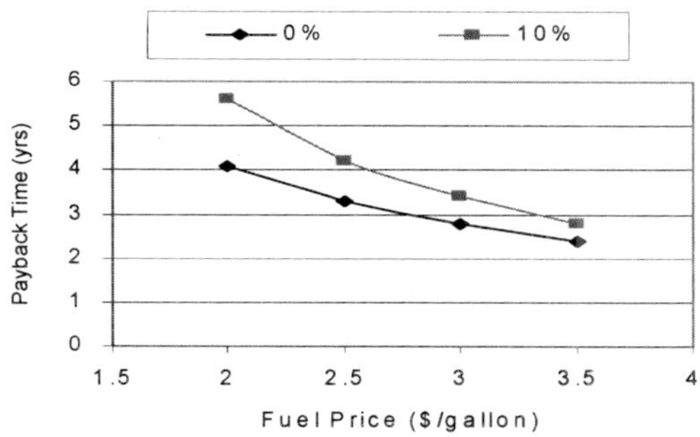

Figure 7: Payback time with respect to fuel price and interest rate (engine with an operation of 8 hours per day).

Acknowledgement

The authors are grateful to US-DOE-NETL for supporting this project (contract number DE-FC26-01NT41248). Thanks are also due to ICRC Corporation for providing the diesel generator set used for the experiment and AVEC Corporation for providing the village diesel generator operation data used for the load pattern analysis.

References

[1] *Alaska Electric Power Statistics*, Institute of Social and Economic Research, University of Alaska Anchorage, Prepared for Alaska Energy Authority, November 2003.
[2] H. T. El-Dessouky, "Fundamentals of Salt Water Desalination," Elsevier Science Ltd, 2002.
[3] J. Dakoozian, "Absorption chilling and ice production," 2004 Alaska Rural energy conference, Session 2B, 2004.
[4] H. Schock, et al., "Thermoelectric conversion of waste heat to electricity in an IC engine powered vehicle," 2007 Diesel Engine-Efficiency and Emissions Research Conference, Aug. 2007.
[5] C. T. Vuk, "Electric turbo compounding technology update," 2007 Diesel Engine-Efficiency and Emissions Research Conference, Aug. 2007.
[6] C. R. Nelson, "Cumins waste heat recovery," 2007 Diesel Engine-Efficiency and Emissions Research Conference, Aug. 2007.
[7] D. J. Patterson and R. W. Kruiswyk, "An engine system approach to exhaust waste heat recovery," 2007 Diesel Engine-Efficiency and Emissions Research Conference, Aug. 2007.
[8] *Rural Alaska Energy Plan- Cogeneration Chapter*, MAFA, Prepared for Alaska Energy Authority, Dec. 2002.
[9] Ground water data for Alaska, 2003, http://waterdata.usgs.fov/ak/nwis/gw.
[10] R. Chellini, "Diesel & Gas Turbine Worldwide," Sept. 2007.
[11] S. Stassel, "An approach to heat recovery in rural Alaska," Alaska Rural Energy Conference, 2004.
[12] A. Telang, "Testing of syntroleum fuels in diesel power plants suitable for Alaska," M.S. thesis, Dept. Mech. Eng., Univ. of Alaska Fairbanks, Fairbanks, Alaska, 2005.
[13] P. R. Raghupatruni, "Performance Analysis of capture of heat energy from diesel engine exhaust," M.S. thesis, Dept. Mech. Eng., Univ. of Alaska Fairbanks, Fairbanks, Alaska, 2007.
[14] R. K. Shah, *Fundamental of Heat Exchanger Design*, John Wiley & Sons, 2003.
[15] V. Ganapathy, *Minimizing Acid Condensation Concerns*, http://vganapathy.tripod.com/corros.html.
[16] *ASHRAE Handbook-Fundamentals*, ASHRAE, 2005.
[17] S. Kanthikiran, "Properties and performance evaluation of syntroleum Synthetic diesel fuels," M.S. thesis, Dept. Mech. Eng., Univ. of Alaska Fairbanks, Fairbanks, Alaska, 2005.
[18] J. Grillot and G. Icart, "Fouling of a cylindrical probe and a finned tube bundle in a diesel exhaust environment," Experimental Thermal and Fluid Science, Vol. 14, Issue 4, 1988, pp 442–454.
[19] W. J. Marner, "Progress in gas-side fouling of heat-transfer surfaces," Applied Mechanics Reviews, Vol. 43, No. 3, March 1990, pp 35–66.
[20] B. Adrian, *Heat Transfer*, John Wiley, 1993.

Heat recovery with low temperature spray drying for thermochemical hydrogen production

V. N. Daggupati, G. F. Naterer & K. S. Gabriel
University of Ontario Institute of Technology, Oshawa, Ontario, Canada

Abstract

This paper examines the heat transfer and evaporative drying of molten salt droplets in a thermochemical copper-chlorine (Cu-Cl) cycle of hydrogen production. An aqueous $CuCl_2$ stream exiting from an electrochemical cell is preheated to 150°C, before entering the flash dryer to produce solid $CuCl_2(s)$. The device must add sufficient heat to remove water and recover solid $CuCl_2$. New innovations of heat recovery aim to develop alternatives to reduce costs and improve efficiency of the evaporation process for $CuCl_2$ particle production. This includes a new method to pressurize the liquid stream sufficiently to atomize droplets through a pressure-reducing nozzle in the spray system, thereby enhancing the device's performance. The liquid phase flashes due to a sudden pressure drop. Unlike other thermochemical cycles, the Cu-Cl cycle can take unique advantage of low-grade waste heat for spray drying and vaporizing processes at low temperatures. The powder produced by spray drying is then injected in a fluidized reactor to eventually produce hydrogen. A common byproduct of manufacturing processes, incinerators, industrial furnaces, etc. – heat – can become a valuable resource through its utility for hydrogen production with the Cu-Cl cycle.
Keywords: heat recovery, hydrogen production, thermochemical Cu-Cl cycle.

1 Introduction

Hydrogen is a promising clean energy carrier of the future, since its oxidation does not emit greenhouse gases that contribute to climate change. A UOIT-led initiative in collaboration with Atomic Energy of Canada Ltd. and other partners is developing a new route to hydrogen production based on advanced CANDU

WIT Transactions on Engineering Sciences, Vol 61, © 2008 WIT Press
www.witpress.com, ISSN 1743-3533 (on-line)
doi:10.2495/HT080101

(CANada Deuterium Uranium) and SCWR (Super Critical Water Reactor) nuclear technologies. Thermochemical water decomposition driven by nuclear heat with a copper-chlorine (Cu-Cl) cycle splits water into hydrogen and oxygen through intermediate copper and chlorine compounds. These chemical reactions form a closed internal loop that re-cycles all chemicals on a continuous basis, without emitting any greenhouse gases externally to the atmosphere. Unlike other sustainable energy resources (such as solar or wind), nuclear heat used in this thermochemical cycle can enable consistently large-scale production of hydrogen. The Cu-Cl cycle could be potentially coupled with nuclear reactors to achieve significantly higher efficiencies and lower costs of large-scale hydrogen production than any other conventional technology.

Hydrogen demand is expected to increase dramatically over the next decade. About 50 million tons/year of hydrogen are needed worldwide (currently about a $282 billion market). The transition to a "hydrogen economy" in the transportation sector would require a lot more hydrogen. In Canada, Alberta's oil sands need over 770,000 tons/year of hydrogen, which is projected to grow to over 2.8 million tons/year in 2020. All major auto-makers are making significant investments in hydrogen vehicles.

Over 200 cycles have been identified previously to produce hydrogen by thermochemical water decomposition from various heat sources, including nuclear and solar energy [1]. Lewis and Taylor [2] reported that a survey of the open literature between 2000 and 2005 did not reveal any more cycles. Very few have progressed beyond theoretical calculations to working experimental demonstrations that establish scientific and practical feasibility of the thermochemical processes.

Much effort internationally is being focused on the sulfur-iodine (S-I) cycle of thermochemical hydrogen production. It is a leading example that has been scaled up to a pilot plant level, with active work conducted by the Sandia National Laboratory (USA), Japan Atomic Energy Agency (JAEA), CEA (France) and others [3, 4]. JAEA has demonstrated a pilot facility up to 30 litres/hr of hydrogen with the S-I cycle. JAEA aims to complete a large S-I plant to produce 60,000 m^3/hr of hydrogen by 2020. This would be sufficient for about 1 million fuel cell vehicles. Japan has a goal to have 5 million fuel cell vehicles on the road by 2020 [5]. Korea (KAERI Institute) is collaborating with China to produce hydrogen with their HTR-10 reactor. Sadhankar et al. [6] have examined the production of hydrogen with high-temperature nuclear reactors. This paper examines a promising low-temperature alternative called the copper-chlorine (Cu-Cl) cycle. This cycle offers a number of advantages over the S-I cycle, including lower operating temperatures, capability of utilizing low-grade waste heat from nuclear reactors and lower-cost materials. This article investigates the heat requirements of low-temperature steps within the Cu-Cl cycle and presents recent UOIT-led developments to build an integrated lab-scale Cu-Cl cycle, in conjunction with Atomic Energy of Canada Ltd. and other partners.

Within the Cu-Cl cycle, an aqueous $CuCl_2$ stream exiting from an electrochemical cell is preheated to 150°C, before entering the flash dryer to produce solid $CuCl_2(s)$. This paper examines whether lower temperature drying can be accomplished, thereby enabling low-grade waste heat recovery from nuclear power plants for purposes of hydrogen production. Several past studies have shown that low-grade heat can be utilized for evaporative drying processes. Lin and Chen [7] used an experimental glass filament method to examine air drying at low temperatures, down to 70 °C with air velocities of 0.55 m/s and an initial droplet diameter of 1.45 mm. A reaction engineering model was used to predict results under conditions of high humidity. A wider range of air velocities between 0.45 and 1.0 m/s was also examined experimentally by the authors [8]. Using a lab-scale spray dryer, Ambike et al. [9] studied an amorphous solid dispersion process with a lab-scale spray dryer, with applications to pharmaceutical research. It was shown that spray drying could be accomplished down to temperatures of 35 °C, with feed rates between 4 – 6 mL/min and outlet air temperatures of 26 °C. Crafton and Black [10] studied n-hexane droplet evaporation at 60 °C on heated horizontal surfaces. Another example of low-temperature spray drying is Walton's study [11] of droplet evaporation with air temperatures down to 23 °C. Farid [12] determined the drying time of a 200 micron droplet with 20-30% solid composition is about 6 seconds at an air temperature of 70 °C. Lin and Gentry [13] studied the effects of air temperature and concentration on drying times of calcium acetate, sodium chloride, ammonium chloride and other inorganic solutions. These past studies provide a useful basis, from which low-temperature spray drying in the Cu-Cl cycle will be examined in this paper.

2 Overview of the thermochemical Cu-Cl Cycle

The Cu-Cl thermochemical cycle uses a series of reactions to achieve the overall splitting of water into hydrogen and oxygen as follows:

$$H_2O(l) \rightarrow H_2(g) + \frac{1}{2}O_2(g) \quad (1)$$

The Cu-Cl cycle splits water into hydrogen and oxygen through intermediate copper and chlorine compounds. These chemical reactions form a closed internal loop that re-cycles all chemicals on a continuous basis, without emitting any greenhouse gases externally to the atmosphere. Steps in the Cu-Cl cycle with and a possible realization of the cycle are shown in fig. 1 and table 1, respectively.

In Step 3 (flash dryer) of the Cu-Cl cycle, an aqueous $CuCl_2$ stream exiting from the electrochemical cell is preheated to 150°C, before entering the flash dryer to produce solid $CuCl_2(s)$, which is required for Step 4. The device must add sufficient heat to remove the water and recover solid $CuCl_2$. New innovations of heat recovery aim to develop alternatives to reduce costs and improve efficiency of the evaporation process for $CuCl_2$ particle production.

3 Evaporative spray drying in the Cu-Cl cycle

The droplet evaporation and drying time depend on the droplet size, air velocity, air temperature, air humidity and operating pressure. Either an energy balance or mass balance method below can be used to estimate the drying time of droplets.

For the former method (energy balance), only three variables are used, namely droplet size, air velocity and temperature, while the latter method (mass transfer) includes all of the five variables. In the analysis below, the following range of variables will be examined: Droplet size: 10–5000 μm, inlet air humidity: 0.0025–0.015 kg water/kg dry air, air temperature: 35–70°C, air velocity: 0.5–4 m/s, operating Pressure: 0.5-1 atm.

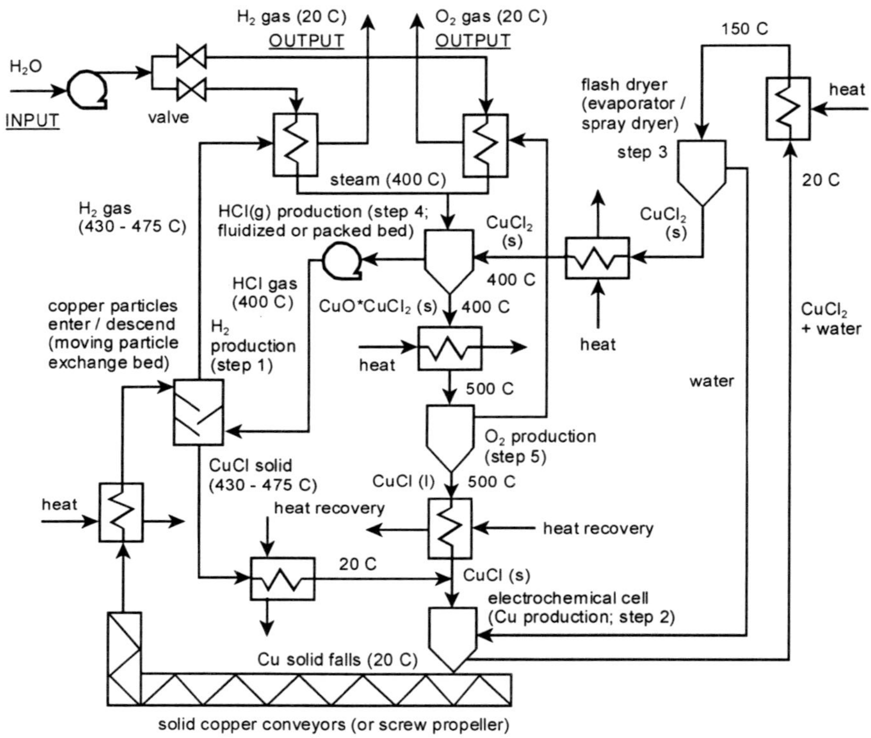

Figure 1: Conceptual layout of the Cu-Cl cycle.

Table 1: Reaction steps in the Cu-Cl thermochemical cycle for hydrogen production.

Step	Reaction	Temperature (°C)
1	$2Cu(s) + 2HCl(g) \rightarrow 2CuCl(l) + H_2(g)$	430 – 475
2	$2CuCl(s) \rightarrow CuCl_2(aq) + Cu(s)$	Ambient (electrolysis)
3	$CuCl_2(aq) \rightarrow CuCl_2(s)$	>100
4	$2CuCl_2(s) + H_2O(g) \rightarrow CuO*CuCl_2(s) + 2HCl(g)$	400
5	$CuO*CuCl_2(s) \rightarrow 2CuCl(l) + 1/2O_2(g)$	500

The evaporation rate of a droplet (kg water / s) is calculated from

$$N_c = \frac{h_c A(\Delta T)}{\lambda} \tag{2}$$

The drying time is then given by

$$t = \frac{\pi d^3 \rho_w}{6 N_c} \tag{3}$$

The convective heat transfer coefficient can be determined from the following correlation for the Nusselt number

$$Nu = \frac{h_c d}{K_d} = 2.0 + 0.6\left(\frac{dV\rho}{\mu}\right)^{0.5}\left(\frac{C_p \mu}{K_d}\right)^{0.33} \tag{4}$$

The mass transfer rate (kg water / s) from the droplet surface to the bulk air stream is calculated as follows,

$$N_c = k_c A(C_s - C_g) M_w \tag{5}$$

The mass transfer coefficient, k_c, can be estimated from the following correlation (6) and the drying time again is calculated using equation (3)

$$Sh = \frac{k_c d}{D_g} = 2.0 + 0.6\left(\frac{dV\rho}{\mu}\right)^{0.5}\left(\frac{\mu}{\rho D_g}\right)^{0.33} \tag{6}$$

From these correlations, and the following figures 2 & 3 and table 2, it can be observed that the drying time rises with an increase in droplet size, inlet air humidity and operating pressure. The drying time decreases with an increase in air temperature and air velocity.

The spray drier volume depends on the air inlet temperature, outlet temperatures, solid concentration, feed flow rates, drying time, humidity and type of spraying. The effects of the above parameters are shown in figure 4 for a hydrogen capacity of 0.241–0.483 kg/day at the outlet air humidity of 90% saturation.

Table 2: Effect of droplet size on drying time at an air temperature of 35°C, humidity of 0.015 kg water/kg dry air, air velocity of 1m/s and drop temperature of 20°C.

Droplet size	10 μm	50 μm	100 μm	300 μm	400 μm	500 μm	1000 μm	3000 μm	5000 μm
Drying time, s	0.03	0.60	2.13	14.95	24.61	36.10	116.4	705.9	1601.1

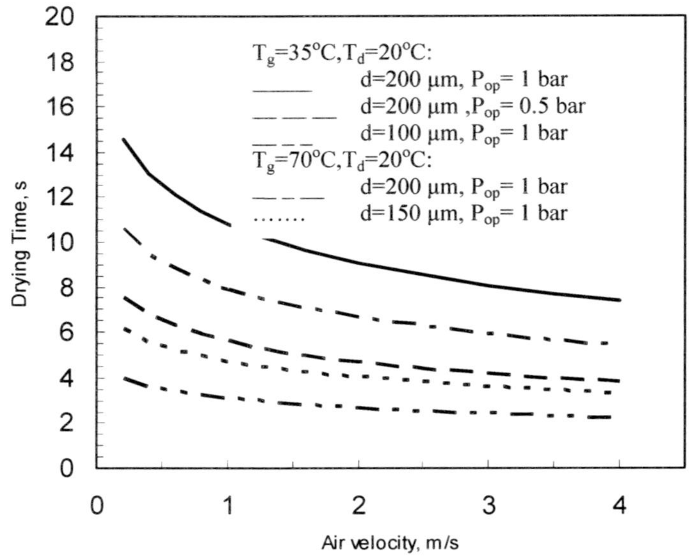

Figure 2: Variation of drying time with air velocity, temperature and operating pressure at an air humidity of H = 0.01 kg water/kg dry air.

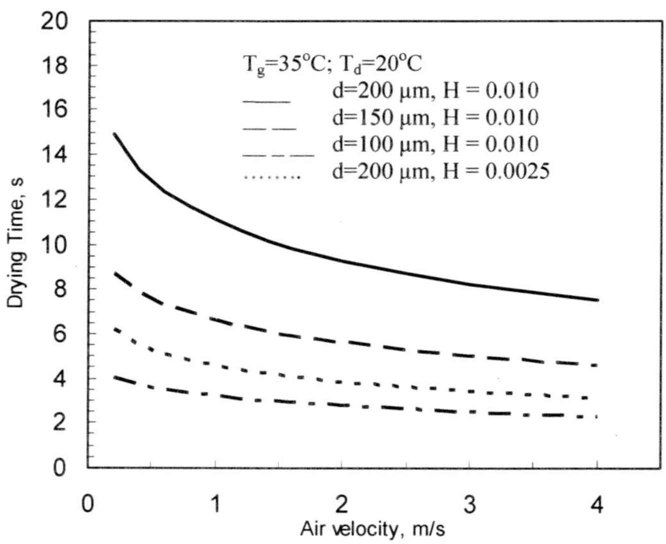

Figure 3: Variation of drying time with air velocity, humidity and drop size.

In a spray drying operation, the selection of inlet and outlet air temperatures is very important. The temperature of the air flow does not need to be higher than the boiling point of water to evaporate the individual drops during the short residence times. The diffusive gradient between the wet surface and unsaturated air leads to evaporation at low temperatures. The optimal choice for the temperature difference between the inlet and the outlet temperatures is the melting point or decay temperature of the product, throughput and final moisture content of the product. The outlet temperature depends on the inlet temperature of air, mass flow rate of air, mass flow rate and solid concentration of the feed.

For small moisture content in the final product, the inlet temperature must be as high as possible and the temperature difference between inlet and outlet air must be as small as possible. Increasing the temperature difference between the inlet and outlet air, while holding the inlet air temperature constant, increases the moisture content in the final product. Figures 2 and 3 show the variation of drying time with air velocity. It was observed that the drying time strongly depends on the inlet air humidity. At a low humidity of 0.0025 kg water/kg dry air, the drying time is less than 6 s for drop sizes less than 200 μm. At a humidity of 0.01 kg water/kg dry air, the drying time is less than 6 s for drop sizes less than 100 μm, at 35°C, and drop sizes less than 150μm, at 70°C. The analysis indicates that evaporative drying is possible down to low temperatures of 35 °C, although low temperature drying may limit the product quality and throughput.

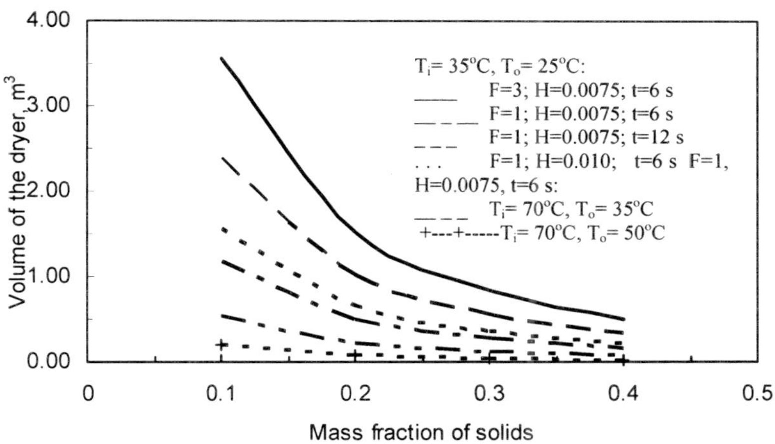

Figure 4: Variation of spray dryer volume with mass fraction of solids, feed flow rate, drying time, inlet air humidity, inlet and outlet air temperature.

For equipment selection in the spraying process of the Cu-Cl cycle, the type of feed material is aqueous cupric chloride with 10-40% (w/w) solid. Commonly used industrial equipment for such processes is a drum dryer or spray dryer. Both dryers can handle liquids and slurries. Also, both dryers have short residence times. Drum dryers produce flakes, while spray dryers produce porous, small rounded particles that are often preferable to flakes. Spray dryers are capable of large evaporation rates. Dust control is intrinsic to spray dryer construction, but not drum driers. A completely enclosed operation of spray dryers also is an advantage when toxic or noxious chemicals are handled.

The function of the spray dryer unit is to dry liquid $CuCl_2$ solution into solid, by utilizing low-grade waste heat from nuclear or other sources. Spray drying involves evaporation of moisture from an atomized feed by mixing the spray and drying medium. The drying medium is typically air. The drying proceeds until the desired moisture content is reached in the sprayed particles and the product is then separated from air. The operating variables of the spray drier involve the atomizer design (type, rotational speed, nozzle diameter and number of nozzles), feed (flow rate, temperature and concentration) and air supply (flow rate, inlet temperature, outlet temperature and humidity). Scale-up of a commercial spray dryer requires optimization of the atomizer, feed, air conditions, and flow rates to determine the residence time. These factors are currently under investigation for construction of a cupric chloride spray dryer for a lab-scale demonstration of the Cu-Cl cycle at UOIT.

4 Conclusions

In thermal power plants, many hundreds of megawatts of waste heat are typically transferred through condensers and moderator vessels to the environment (normally a nearby lake). This paper has examined low-temperature spray drying in a thermochemical cycle to reduce thermal pollution, particularly how waste heat can be diverted to cupric chloride drying for powder production in the Cu-Cl cycle, rather than heat rejection to a lake. The powder product can then be transported to a fluidized reactor that eventually produces hydrogen. The Cu-Cl cycle is a promising alternative to the sulfur-iodine cycle (S-I cycle; currently pursued by countries such as Japan, USA and France), due to its lower operating temperatures, lower cost materials and waste heat utilization from any nuclear or other industrial heat source(s).

Acknowledgements

Financial support of this research from Atomic Energy of Canada Ltd., Ontario Research Excellence Fund and the Natural Sciences and Research Council of Canada (NSERC) is gratefully acknowledged.

Nomenclature

A droplet surface area, m^2
C_p specific heat, J/kgK
C_s vapor concentration at the droplet surface, kmol/ m^3

C_g vapor concentration in the bulk air, kmol/ m^3
D_g, diffusion coefficient for vapor in air, m^2/s
F feed flow rate, kg/hr
d droplet diameter, m
K_d thermal conductivity, W/mK
k_c mass transfer coefficient, m/s
h_c heat transfer coefficient, W/m^2K
H humidity of inlet air, kg water/kg dry air
M_w Molecular weight of water, 18 kg/kmol
P_{sat} vapor pressure of water at drop temperature, N/m^2
P_{op} operating pressure, N/m^2
R gas constant, 8314 Nm/kmolK
t drying time, s
T_d droplet temperature, K
T_g bulk air temperature, K
T_i air inlet temperature, K
T_o air outlet temperature, K
V air velocity, m/s
X_i mole fraction of water vapor in air,
ρ density of air, kg/m^3
ρ_w density of water, kg/m^3
μ viscosity of air, kg/ms
λ latent heat of water, 2417.44 kJ/kg
ΔT temperature difference between drying air and droplet

References

[1] McQuillan, B. W., Brown, L. C., Besenbruch, G. E., Tolman, R., Cramer, T., Russ, B. E., Vermillion, B. A., Earl, B., Hsieh, H. T., Chen, Y., Kwan, K., Diver, R., Siegal, N., Weimer, A., Perkins, C. & Lewandowski, A, High efficiency generation of hydrogen fuels using solar thermochemical splitting of water: *Annual Report*, GA-A24972, General Atomics, San Diego, CA.
[2] Lewis, M. & Taylor, A., High temperature thermochemical processes, DOE Hydrogen Program, *Annual Progress Report*, pp. 182–185, 2006, Washington, DC.
[3] Sakurai, M., Nakajima, H., Amir, R., Onuki, K. & Shimizu, S., Experimental study on side-reaction occurrence condition in the iodine-sulfur thermochemical hydrogen production process *International Journal of Hydrogen Energy*, 23, pp. 613–619, 2000.
[4] Schultz, K., Thermochemical production of hydrogen from solar and nuclear energy, *Technical Report for the Stanford Global Climate and Energy Project*, General Atomics, San Diego, CA, 2003.
[5] Fuel Cell Vehicles: Race to a new automotive future, US Department of Commerce, 2002.

[6] Sadhankar, R. R., Li, J., Li, H., Ryland, D. & Suppiah, S., Hydrogen generation using high-temperature nuclear reactors *55th Canadian Chemical Engineering Conference*, October, Toronto, 2005.

[7] Lin, S. X. & Chen, X. D., Prediction of air-drying of milk droplet under relatively high humidity using the reaction engineering approach, *Drying Technology*, 23, pp. 1395–1406, 2005.

[8] Lin, S. X. & Chen, X. D., Air drying of milk droplet under constant and time dependent conditions, *AIChE Journal*, 51, pp.1790–1799, 2005.

[9] Ambike, A., Mahadik, K. R. & Paradkar, A., Spray dried amorphous solid dispersions of simvastain a low Tg drug: in vitro and in Vivo evaluations, Pharmaceutical *Research*, 22, pp. 990–998, 2005.

[10] Crafton, E. F. & Black, W. Z., Heat transfer and evaporation rates of small liquid droplets on heated horizontal surfaces, *International Journal of Heat and Mass Transfer*, 47, pp. 1187-1200, 2004.

[11] Walton, D. E., The evaporation of water droplets: a single droplet drying experiment, *Drying Technology*, 22, pp.431–456, 2004.

[12] Farid, M., A new approach to modelling of single droplet drying, *Chemical Engineering, Science,* 58 pp. 2985–2993, 2003.

[13] Lin, J. C. & Gentry, J. W., Spray drying drop morphology: experimental study, *Aerosol Science and Technology*, 37, pp. 15–32, 2003.

Section 5
Heat transfer

Peculiarities of heat transfer from in-line tube bundles to upward aqueous foam flow

J. Gylys[1], S. Sinkunas[2], B. Sundén[3], I. Gabrielaitiene[1] & T. Zdankus[1]
[1]*Energy Technology Institute, Kaunas University of Technology, Lithuania*
[2]*Department of Thermal and Nuclear Energy, Kaunas University of Technology, Lithuania*
[3]*Department of Energy Sciences, Lund University, Sweden*

Abstract

Four in-line tube bundles with different geometry were investigated for establishing their performance in terms of heat transfer enhancement. Two-phase aqueous foam was used as a coolant. Such coolant was considered, because our previous research showed that large heat transfer intensity may be reached even at small mass flow rate of the foam. Spacing among the centres of the tubes across the first in-line tube bundle was 0.03 m and spacing along the bundle was 0.03 m. In the second case spacing among the centres of the tubes across the bundle was 0.03 m; spacing along the bundle was 0.06 m. In the third case spacing was accordingly 0.06 and 0.03 and in the last case spacing was accordingly 0.06 m and 0.06 m. During an experimental investigation it was determined a dependence of heat transfer intensity on flow parameters. The investigation of heat transfer from the bundle to upward vertical foam flow was provided for three different values of foam volumetric void fractions β=0.996÷0.998. The velocity of the foam flow was changed from 0.14 to 0.30 m/s. The heat transfer coefficient varied from 200 to 2000 W/(m^{2}K) for the above mentioned foam flow parameters.

Keywords: in-line tube bundle, aqueous foam, upward flow, heat transfer.

1 Introduction

Heat exchangers with different type and geometry of tube bundles are usually used in industry. Two-phase foam coolant has a number of advantages in comparison with single-phase coolant. For example, it requires smaller coolant flow rates, and lower energy consumption for delivering the coolant to the region where heat transfer process takes place. Two-phase foam flow can signally reduce material expenditures. A totally new, modern and economic heat exchanger with simple and safe exploitation using two-phase foam flow in heat transfer could be created. It would be compact, light and with greater intensity of heat transfer.

Aqueous foam flow has number of peculiarities which always exists during heat transfer process: drainage of liquid from foam [1, 2], diffusive transfer of gas between bubbles [3], division and collapse of foam bubbles [4]. Those phenomena are closely related with each other and make an application of analytical study of the heat transfer processes in the foam flow extremely complicated. So, an experimental approach has been selected as the most suitable for our present investigation.

Heat transfer of different tube bundles to one-phase fluid was investigated enough [5, 6], but practically there are no data about tube bundles heat transfer to foam flow. This work is an extension of our previous works [7–9]. Heat transfer from four in-line tube bundles with different types of geometries to the vertical upward statically stable foam flow was investigated experimentally. The influence of in-line tube bundles geometry was analyzed for average heat transfer intensity by increasing the distance between tubes centres of tube. Dependence of heat transfer intensity on flow parameters was determined. Objectives of this investigation are to determine an optimal geometry of in-line tube bundle which guarantees a maximal heat transfer to two-phase foam flow.

The results of investigation were treated analyzing relationships between Nusselt and Reynolds numbers and volumetric void fraction.

2 Experimental set-up

Investigation was performed on the experimental set-up consisting of experimental channel, foam generator, in-line tube bundle, gas and liquid flow meters, liquid storage reservoir, compressor, electric current transformer and stabilizer [7–9]. Cross section of the channel had dimensions (0.14 x 0.14) m^2; height of experimental channel was 1.8 m.

Statically stable foam (one type of aqueous foam) was used for experiments [7–9]. Foam-able liquid for the statically stable foam generation was produced from water solution with the detergents. Concentration of the detergents was kept constant at 0.5% in all experiments. Foam-able liquid was supplied from the reservoir onto the special perforated plate for the foam generation from all channel sides; gas was supplied to the perforated plate from the bottom. Air from the compressor via receiver was used as "gas" in all investigations. Foam flow was produced during gas and liquid contact on the perforated plate.

The perforated plate was made from stainless steel plate with thickness of 2 mm. Holes were located in staggered order. Diameter of the holes was 1 mm; spacing among centres – 5 mm.

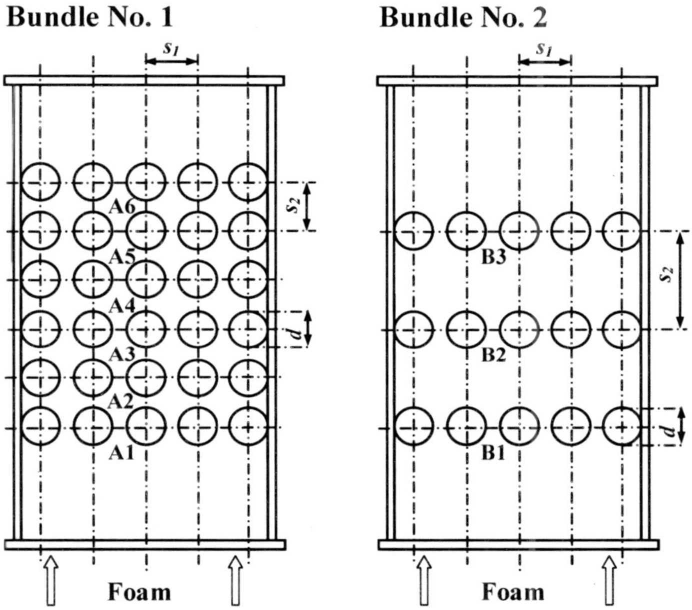

Figure 1: In-line tube bundles No. 1 and No. 2 in upward foam flow.

Four in-line tube bundles were used during experimental investigation. Schematic view of the experimental channel with tube bundles No. 1 and No. 2 is shown in the Fig. 1. The in-line tube bundle No. 1 consisted of five columns with six tubes in each. Spacing between centres of the tubes was s_1=s_2=0.03 m. The in-line tube bundle No. 2 (Fig. 1) consisted of five columns with three tubes in each. Spacing between centres of the tubes across the experimental channel was s_1=0.03 m and spacing along the channel was s_2=0.06 m.

The in-line tube bundle No. 3 (Fig. 2) consisted of three columns with six tubes in each. Spacing between centres of the tubes across the experimental channel was s_1=0.06 m and spacing along the channel was s_2=0.03 m. The in-line tube bundle No. 4 (Fig. 2) consisted of three columns with three tubes in each. Spacing between centres of the tubes was s_1=s_2=0.06 m.

External diameter of all the tubes was equal to 0.02 m. An electrically heated tube – calorimeter had an external diameter equal to 0.02 m also. During the experiments calorimeter was placed instead of one tube of the bundle. An electric current value of heated tube was measured by an ammeter and voltage by a voltmeter. Temperature of the calorimeter surface was measured by eight calibrated thermocouples: six of them were placed around the central part of the calorimeter and two of them were placed in both sides of the calorimeter at a distance of 50 mm from the central part. Temperature of the foam flow was measured by two calibrated thermocouples: one in front of the bundle and one behind it.

Measurement accuracies for flows, temperatures and heat fluxes were of range correspondingly 1.5%, 0.15–0.20% and 0.6–6.0%.

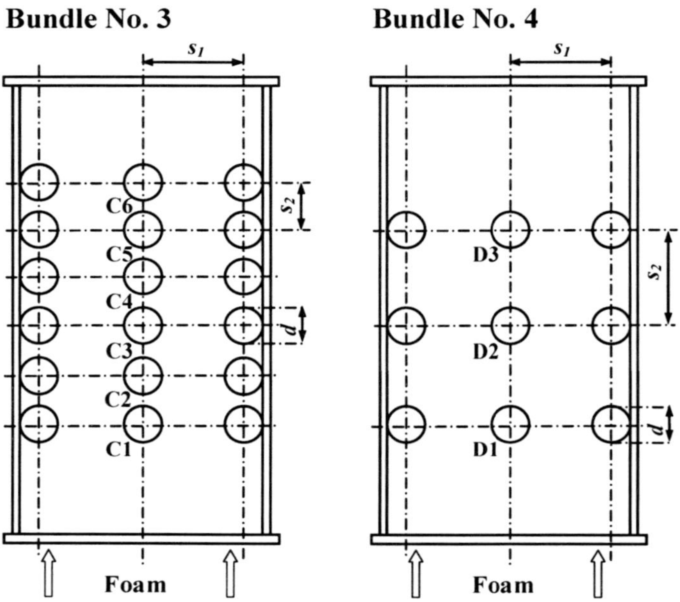

Figure 2: In-line tube bundles No. 3 and No. 4 in upward foam flow.

During the experimental investigation a relationship was obtained between an average heat transfer coefficient h from one side and foam flow volumetric void fraction β and gas flow Reynolds number Re_g from the other side:

$$Nu_f = f(\beta, \mathrm{Re}_g). \tag{1}$$

Nusselt number was computed by formula

$$Nu_f = \frac{hd}{\lambda_f}. \tag{2}$$

Here λ_f is the thermal conductivity of the statically stable foam flow, W/(m·K), computed by the equation

$$\lambda_f = \beta\lambda_g + (1-\beta)\lambda_l. \tag{3}$$

An average heat transfer coefficient we calculated as

$$h = \frac{q_w}{\Delta T}. \tag{4}$$

Gas Reynolds number of foam flow we computed by formula

$$\mathrm{Re}_g = \frac{G_g d}{A \nu_g}. \quad (5)$$

Foam flow volumetric void fraction we expressed by the equation

$$\beta = \frac{G_g}{G_g + G_l}. \quad (6)$$

Experiments we performed within limits of Reynolds number diapason for gas (Re_g): 190–410 (laminar flow regime) and foam volumetric void fraction (β): 0.996–0.998. Gas velocity for foam flow was changed from 0.14 to 0.30 m/s.

3 Results

Heat transfer from the tube bundle No. 1 to the vertical upward foam flow was investigated initially [8, 9]. Then tube bundle No. 2 was installed in an experimental channel instead of the bundle No. 1 and so on.

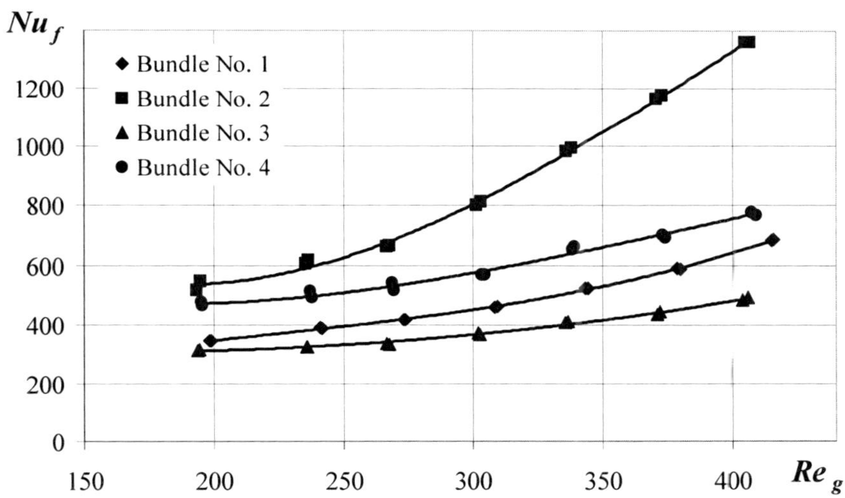

Figure 3: An average heat transfer intensity of the tubes from middle column of the in-line bundle No. 1, No. 2, No. 3 and No. 4 to upward foam flow, β=0.996.

Heat transfer intensity of the tubes located at the different places across and along the in-line tube bundle to vertical upward foam flow was investigated experimentally. In the foam flow case heat transfer intensity of the different tubes is under the influence of distribution of local flow velocity and local foam void fraction across and along the channel [8, 9]. The third factor which influences the heat transfer intensity of tubes is the foam structure.

An average heat transfer rate of tubes from middle column was calculated in order to compare the experimental results of in-line tube bundles with different geometries. An average heat transfer intensity of the tubes from middle column (A, B, C and D) of the in-line bundles No. 1, No. 2, No. 3 and No. 4 to upward foam flow at β=0.996 is shown in Fig. 3.

The effect of "shadow" takes place in the case of reduced spacing between centres of the tubes in the columns (tube bundles No. 1 and No. 3). Therefore an average heat transfer intensity of the tubes from middle column of the bundles No. 2 and No. 4 is higher than that of the bundles No. 1 and No. 3.

Local foam flow velocity increases while foam passes the rows of tubes. Local velocity is higher for the tube rows with more tubes in each (tube bundles No. 1 and No. 2). Therefore an average heat transfer intensity of the tubes from middle column of the bundles No. 1 and No. 2 is accordingly higher than that of the bundles No. 3 and No. 4.

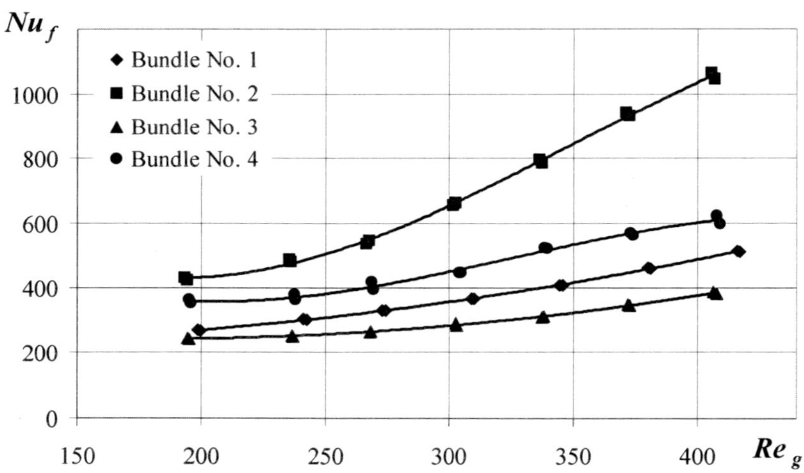

Figure 4: An average heat transfer intensity of the tubes from middle column of the in-line bundle No. 1, No. 2, No. 3 and No. 4 to upward foam flow, β=0.997.

By increasing of foam flow gas Reynolds number (Re_g) from 190 to 410, heat transfer intensity (Nu_f) of the tubes from middle column to upward foam flow increases twice of the in-line bundle No. 1, by 2.5 times of the bundle No. 2 and by 1.6 times of the bundles No. 3 and No. 4 for β=0.996. An average heat transfer intensity of the tubes from middle column of the bundle No. 2 is on average by 1.8 times higher than that of the bundle No. 1, by 2.3 times higher than that of the bundle No. 3 and by 1.5 times higher than that of the bundle No. 4 for β=0.996 and Re_g=190÷410.

An average heat transfer intensity of the tubes from middle column (A, B, C and D) of the in-line tube bundles No. 1, No. 2, No. 3 and No. 4 to upward foam flow at β=0.997 is shown in Fig. 4. In this case foam is drier and heat transfer intensity of all investigated bundles is less in comparison with foam volumetric void fraction (β) equal to 0.996. Changing Re_g from 190 to 410, an average heat transfer intensity of the tubes from middle column to upward foam flow increases by 1.9 times of in-line bundle No. 1, by 2.5 times of the bundle No. 2, by 1.6 times of the bundle No. 3 and by 1.7 times of the bundle No. 4 for β=0.997. An average heat transfer intensity of the tubes from middle column of the bundle No. 2 is on average by 1.9 times higher than that of the bundle No. 1, by 2.3 times higher than that of the bundle No. 3 and by 1.5 times higher than that of the bundle No. 4 for β=0.997 and Re_g=190–410.

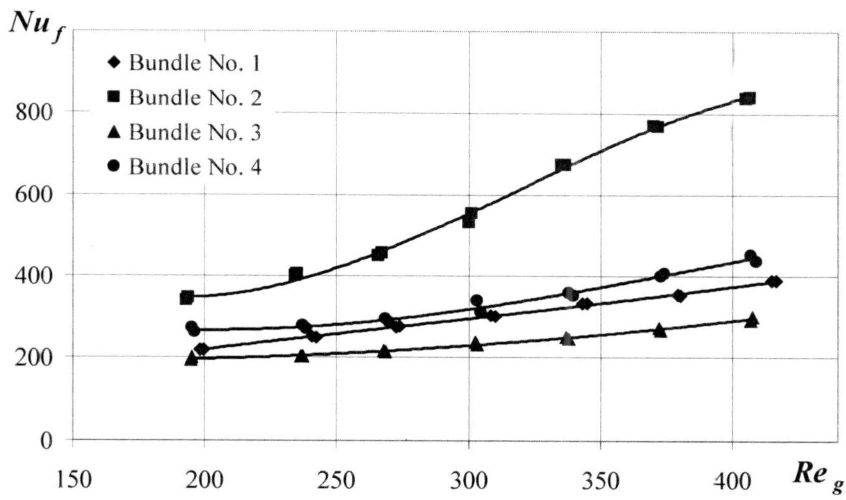

Figure 5: An average heat transfer intensity of the tubes from middle column of the in-line bundle No. 1, No. 2, No. 3 and No. 4 to upward foam flow, β=0.998.

An average heat transfer rate of the tubes from middle column of the in-line tube bundles No. 1, No. 2, No. 3 and No. 4 to upward foam flow at β=0.998 (the driest foam of our investigation) is shown in Fig. 5. With increase of Re_g from 190 to 410, an average heat transfer intensity of the tubes from middle column to upward foam flow increases by 1.8 times of in-line bundle No. 1, by 2.4 times of the bundle No. 2, by 1.5 times of the bundle No. 3 and by 1.7 times of the bundle No. 4 for β=0.998. An average heat transfer intensity of the tubes from middle column of the bundle No. 2 is on average by 1.9 times higher than that of the bundle No. 1, by 2.4 times higher than that of the bundle No. 3 and by 1.7 times higher than that of the bundle No. 4 for β=0.998 and Re_g=190–410.

Experimental results of investigation of heat transfer from the in-line tube bundles to upward laminar statically stable foam flow were generalized by criterion equation using dependence between Nusselt number Nu_f and gas Reynolds Re_g number. This dependence within the interval $190 < Re_g < 410$ for the in-line tube bundle in upward foam flow with the volumetric void fraction β=0.996, 0.997, and 0.998 can be expressed as follows:

$$Nu_f = c\beta^n \mathrm{Re}_g^m . \tag{7}$$

On average, for whole middle column (A) of the in-line tube bundle No. 1 (s_1=s_2=0.03 m) to the upward foam flow c=5.7, n=340, m=102.1(1.006–β).

On average, for whole middle column (B) of the bundle No. 2 (s_1=0.03 and s_2=0.06 m) to the upward foam flow c=0.29, n= –125, m=14.3(1.089–β).

On average, for whole middle column (C) of the bundle No. 3 (s_1=0.06 and s_2=0.03 m) to the upward foam flow c=7.4, n= –111, m=22.8(1.023–β).

On average, for whole middle column (D) of the bundle No. 4 (s_1=0.06 and s_2=0.06 m) to the upward foam flow c=25.4, n=363, m=77.5(1.006–β).

4 Conclusions

Heat transfer from four in-line tube bundles with different geometry to vertical laminar upward statically stable foam flow was investigated experimentally.

Heat transfer intensity of in-line tube bundles to upward foam flow is higher for bundles with increased spacing between tubes centres along the bundle (bundles No. 2 and No. 4).

Local foam flow velocity increases while foam passes the rows of tubes and influences the heat transfer intensity of tubes. Therefore heat transfer intensity is higher for the tube bundles with reduced spacing between tubes centres across the bundle (bundles No. 1 and No. 2).

Criterion equation (7) may be applied for calculation and design of the statically stable foam heat exchangers with in-line tube bundles.

Nomenclature

A – cross section area of experimental channel, m^2; c, m, n – coefficients; d – outside diameter of tube, m; G – volumetric flow rate, m^3/s; Nu– Nusselt number; q – heat flux density, W/m^2; Re – Reynolds number; $\overline{T}$ – average temperature, K; h – average coefficient of heat transfer, $W/(m^2 \cdot K)$; β – volumetric void fraction; λ – thermal conductivity, $W/(m \cdot K)$; ν – kinematic viscosity, m^2/s.

Indexes

f – foam;
g – gas;
l – liquid;
w – wall of heated tube.

Acknowledgement

The authors gratefully acknowledge the financial support of the Lithuanian Science Foundation.

References

[1] Gylys, J., Hydrodynamics and Heat Transfer Under the Cellular Foam Systems, Technologija: Kaunas, 1998.

[2] Fournel B., Lemonnier H., Pouvreau J., Foam Drainage Characterization by Using Impedance Methods, 3rd Int. Symp. on Two-Phase Flow Modelling and Experimentation, Pisa, Italy, p. [1–7], 2004.

[3] Tichomirov V., Foams. Theory and Practice of Foam Generation and Destruction, Chimija: Moscow, 1983.

[4] Garrett P. R, Recent developments in the understanding of foam generation and stability. Chemical Engineering Science, **48(2)**, pp. 367–392, 1993.

[5] Zukauskas A., Convectional Heat Transfer in Heat Exchangers, Nauka: Moscow, p. 472, 1982.

[6] Hewitt, G. F., Heat exchanger design handbook 2002, York, Begell House, 2002.

[7] Gylys J., Zdankus T., Sinkunas S., Giedraitis V., Study of In-line Tube Bundle Cooling in Vertical Foam Flow. WSEAS Transactions on Heat and Mass Transfer, Iss. 6, **1**, pp. 632–637, 2006.

[8] Gylys, J.; Sinkunas, S.; Zdankus, T.; Giedraitis, V., Different type tube bundle heat transfer to vertical foam flow, Proceedings of the 5th International Conference on Nanochannels, Microchannels and Minichannels, ICNMM 2007, Pueble, Mexico, pp. 449–455, 2007.

[9] Gylys, J.; Sinkunas, S.; Zdankus, T.; Giedraitis, V., Balcius, A., Foam flow turn influence on the in-line tube bundle heat transfer intensity, Computational Methods and Experimental Measurements XIII, Prague, Czech Republic, pp. 457–464, 2007.

Heat transfer in reactive Co/Al nanolaminates

M. L. Hobbs[1], D. P. Adams[2] & J. P. McDonald[2]
[1]*Nanoscale & Reactive Processes, Sandia National Laboratories, USA*
[2]*Thin Film, Vacuum & Packaging, Sandia National Laboratories, USA*

Abstract

Reaction front propagation rates of free standing multilayer thin foils of Co/Al have been determined using a diffusion limited reaction model by means of a method-of-lines code with a stiff solver and adaptive gridding. Predicted and measured reaction front speed variations with bilayer thickness, t_b, can be separated into three regimes. The three regimes are delineated by the critical bilayer thickness, $t_{b,c}$, and the bilayer thickness that produces the maximum front velocity, $t_{b,max}$. The critical bilayer is composed predominately of premixed or fully reacted Co_xAl_y with a thickness of ~2.7 nm. The front velocity in the three regimes 1) is zero when $0 \leq t_b \leq t_{b,c}$ since there are no reactants 2) increases when $t_{b,c} < t_b < t_{b,max}$ since the reactant concentration increases with t_b and 3) decreases when $t_{b,max} < t_b$ since the diffusive resistance increases with t_b. The sensitivity of front velocity to property variation is discussed. Steady and oscillatory combustion are predicted for this material pair.
Keywords: Co/Al, gasless metal combustion, diffusion, nanolaminate.

1 Introduction

Alternating nanoscale layers of pure metals can be ignited to produce localized and rapid heating for improved soldering or brazing of microelectronic assemblies [1]. Exothermic metal multilayers used for these applications consist of hundreds or thousands of individual reactant layers generally having a single out-of-plane periodicity. Figure 1.A shows a cross section of a multilayer nanolaminate of Co and Al deposited by direct current sputtering [2]. The layer periodicity is uniform through thickness and the metal layers are high-purity, polycrystalline metal. As shown in Figure 1.B, a 27 Å thick premixed Co_xAl_y interlayer forms at the interfaces between the Co and Al layers.

WIT Transactions on Engineering Sciences, Vol 61, © 2008 WIT Press
www.witpress.com, ISSN 1743-3533 (on-line)
doi:10.2495/HT080121

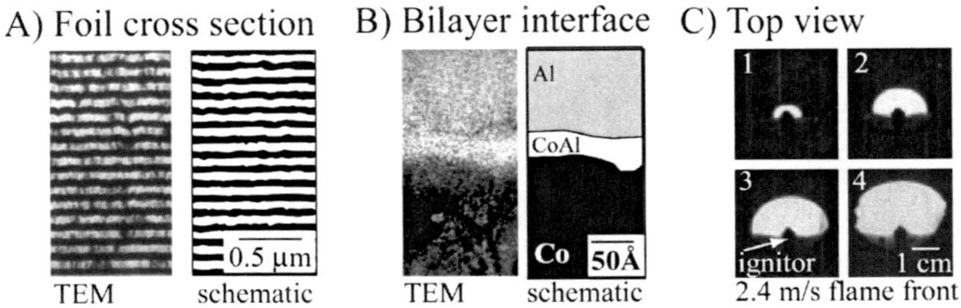

Figure 1: A) Example foil cross section showing multiple bilayers, B) evidence of amorphous Co_xAl_y layer between the Co and Al layers, and C) plan view of foil during reaction. Transmission electron microscopy (TEM) was used for A and B and high-speed photography in C.

Foils of various design exhibit self-propagating high temperature synthesis. Reaction occurs whether ignited in air (as part of the current study) or in vacuum. Figure 1.C shows a plan view of a thin multilayer nanolaminate at various times after single point ignition. Detached foils were ignited over the apertures of metal washers. The reacted portion of the foils is lighter in color than the unreacted portion.

In this proceeding, we model the dynamics of self-propagating high temperature synthesis as a function of multilayer design. In general, the trends in the current work are consistent with previous work involving other exothermic material pairs [3]. All of the foils are free standing in the current work. The behaviour of nanolaminates reacted on a substrate is discussed in reference [4].

2 Diffusion limited reaction model

Hardt and Phung [5] developed an analytical model for propagation of gasless reactions in solids. This model was developed for binary metal powder mixtures by assuming that the metals were one dimensional (layered) and that the thermophysical properties were independent of temperature. Hardt and Phung further assumed that melting of the reactants do not affect the burning process and that the specimen is large compared to the front width. These assumptions imply that heat loss by radiation or convection from the surface of the specimen does not disturb the internal temperature in or near the reaction zone.

In the current work, reactive foils are modelled as a continuum since the dimensions of the foils are too small to model discretely. For example, to discretize the thickness of the mixed Co_xAl_y layer shown in Figure 1.B with three elements would require billions of elements to model the complete system shown in Figure 1.C, even as a 2D axisymmetric system. Furthermore the time steps would be prohibitively small. The continuum assumption allows the reactive nanolaminates to be modelled with a constitutive model similar to the model described by Hardt and Phung [5].

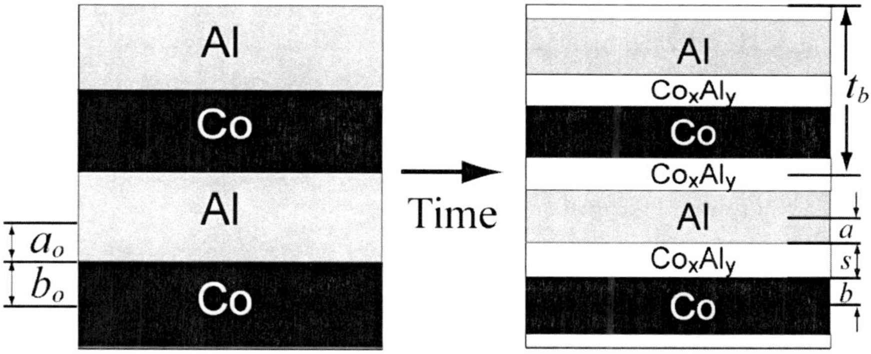

Figure 2: Two bilayers of Al and Co showing initial half thicknesses a_o and b_o, with $t_b = 2(a + b + s)$. After time a mixed layer s with composition Co_xAl_y is formed between the Al and Co layers.

Figure 2 shows a cross section of two bilayers with a, b, and s representing the half thickness of phase A and B, and the full thickness of the mixed layer s. For example, A represents aluminium, B represents cobalt, and s represents the mixed or fully reacted layer Co_xAl_y. The subscript "o" denotes the layer thicknesses before the initial interfacial layer is formed. The "initial" interfacial layer thickness is represented by s_i and is assumed to be constant. The initial thicknesses of the material A and B are determined once the initial thickness of the reacted layer, s_i, is specified.

A one-dimensional (1D), multiple layer, method-of-lines temperature based chemistry code, Tchem, has been developed for spherical, cylindrical, or slab geometries with a general library for inert and reactive nanolaminate materials. Tchem is based on Sandia National Laboratories' explosive chemistry code Xchem [6]. Tchem includes adaptive gridding, multiple phase changes, volumetric source terms, enclosure radiation, and contact resistance. The energy equation solved by Tchem is $\rho c(\partial T/\partial t)+\nabla\cdot q=-Q\rho(\partial F/\partial t)$, where ρ, c, T, q, Q, and F represent material density, specific heat, temperature, conductive energy flux using Fourier law ($q=-k\nabla T$, with k being thermal conductivity), energy release, and unreacted fraction. Table 1 summarizes the equations used in the diffusion model. An algorithm for implementing the model is given in the footnotes of Table 1. The algorithm accounts for the initial mixed "s_i" layer that forms before ignition. Table 2 gives the parameters for the diffusion model. Table 3 gives thermophysical properties for the Co/Al system discussed in the current work. The properties of the intermetallic were determined with common mixture rules, e.g. the specific heat is mass fraction weighted and the thermal conductivity is volume fraction weighted.

The bilayer material is partially melted as the temperature increases to the melting point of Al. The remainder of the bilayer material liquefies when the melting point of the cobalt is exceeded, although not all designs reach this temperature. The bilayer material no longer is composed of separate materials

Table 1: Auxiliary equations used in the continuum diffusion model[†].

a	Thickness of material A	$a = \dfrac{a_o + b_o - s}{1 - \dfrac{\rho_A}{\rho_B}\dfrac{M_{w,B}}{M_{w,A}}\dfrac{N_B}{N_A}}$	(1)
b	Thickness of material B	$b = a\dfrac{\rho_A}{\rho_B}\dfrac{M_{w,B}}{M_{w,A}}\dfrac{N_B}{N_A}$	(2)
s	Thickness of reacted layer	$\dfrac{ds}{dt} = \dfrac{Dw}{s}$, where $w = (1 + b_o / a_o)$	(3)
F	Unreacted fraction	$F = 1 - \dfrac{s}{a_o w}$ and $\dfrac{dF}{dt} = -\dfrac{D}{a_o s}$	(4)
ρ	Density of multi-layer foil	$\rho = \rho_{AB} = \dfrac{a_o\rho_A + b_o\rho_B}{a_o + b_o}$	(5)
D	Diffusion coefficient	$D = D_o \exp\left(-\dfrac{E}{RT}\right)$	(6)

[†]Algorithm for implementing model:

a) Choose a_o, calculate b_o using (2).
b) Choose s_i (this is the *initial* premixed layer that forms prior to ignition).
c) Calculate a_i using s_i and (1).
d) Calculate b_i using (2).
e) Calculate F_i using s_i and (4).
f) Solve ordinary differential equations in (3) and (4) for s and F.

Table 2: Diffusion model parameters.

Property	Description	Co/Al*
s_i, m	Initial mixed layer thickness	2.5×10^{-9}-2.9×10^{-9}
D_o, m^2/s	Initial diffusion coefficient	4.5×10^{-10}-5.5×10^{-9}
E/R, K	Activation Energy/gas constant	5035.8

*Ranges are assumed to be uniformly distributed.

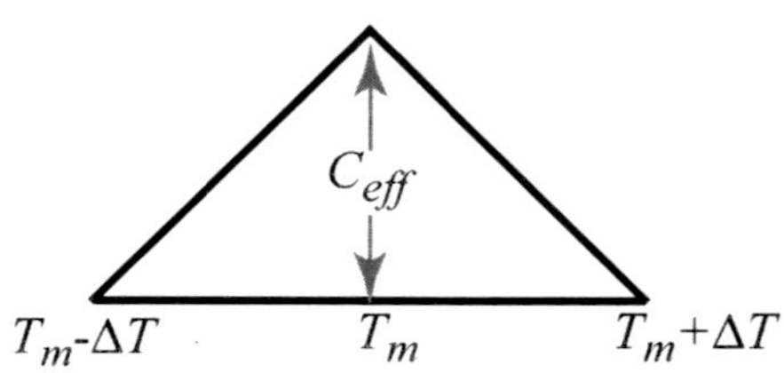

$Area = ½(base \times height)$
$h_{fus} = ½(2\Delta T \times C_{eff})$
$C_{eff} = h_{fus}/\Delta T$

Figure 3: The area of the triangle is the latent enthalpy of fusion. The effective heat capacity is the height of the triangle with the base of the triangle equal to the temperature interval of the mush zone, e.g. 2ΔT.

once the material has reacted, e.g. $F = 0$. Solidification of the intermetallic occurs in two steps using the dual enthalpies of fusion in Table 3. Phase change is modelled using an effective capacitance model [7] with a mush zone as shown in Figure 3. The effective heat capacity at the melting point is the latent enthalpy divided by ΔT as shown in Figure 3. The specific heat is held constant until $T_m - \Delta T$. The specific heat is then increased linearly with temperature to a value of $c + c_{eff}$ at T_m. The specific heat is then decreased linearly to a value of c at $T_m - \Delta T$. The effective capacitance model is used for both phase changes.

Table 3: Thermophysical properties of Al, Co, and Co/Al[a].

Property	Al	Co	Co/Al
c, J/kgK	888–906	417–425	565–576
h_f, J/g	0	0	-1,260 to -1,400
h_{fus}, J/g	397	273	125, 187
k, W/mK	213–261	90–110	164–201
M_w, g/mol	27.0	58.9	85.9
ρ, g/cm^3	2.70	8.86	5.16
T_m, K	934	1770	934, 1770

[a]All properties except heats of formations, h_f, are from reference [8]. Heats of formation are from reference [9]. Property ranges are assumed to be uniformly distributed. These ranges are used with the uncertainty predictions.
[b]The Co/Al foil is assumed to partially melt at the melting point of both the pure metals. The two latent enthalpies for the Co/Al foil are calculated as: $h_{fus,Co/Al} = h_{fus,Al}(M_{w,Al} / M_{w,AlCo})$ and $h_{fus,Co}(M_{w,Co} / M_{w,AlCo})$.

An efficient *constrained sampling* technique known as Latin hypercube sampling (LHS) was used to determine the sensitivity of the predicted front velocities to uncertainty in the eight (8) input parameters: k_{Al}, c_{Al}, D_o, s_o, k_{Co}, c_{Co}, ΔT, and h_{rzn}. All of these parameters except ΔT are listed in Tables 2 and 3 with an assumed uniform distribution. ΔT is assumed to vary uniformly from 4-5 K. The LHS technique developed by McKay et al. [10] selects n different values for each of the 8 uncertain variables. In the current work, the number of samples, n, was chosen to be 9. The range of each input parameter was divided into n nonoverlapping intervals based on equal probability. One random value from each interval was selected according to a uniform probability density function in the interval. The n values were then paired in a random manner with the n values of the other parameters. The front velocities were then calculated n times with the n different sets of input parameters. The mean and standard deviation of the velocities were then calculated from the n sets of responses.

The relative importance of each model parameter to the uncertainty in the predicted front velocity was calculated by fitting a simple model to the 9 LHS runs, $v = A_0 + A_1 k_{Al} + ... + A_8 E / R$, and using a mean value analysis to determine the importance. The coefficients, A_i, were determined by setting up 9 equations with 9 unknown coefficients and solving for the coefficients. The derivatives of V, with respect to input parameters i are the sensitivity coefficients, which are also the regression coefficients, *e.g.* $\partial V/\partial k_{Al} = A_1$. The variance of the velocity, sensitivity coefficients, and importance factors are then:

$$\sigma_V^2 = \sum_{i=1}^{8}(\sigma_i A_i)^2 \text{ , } \gamma_i = \frac{\sigma_i}{\sigma_T} \times A_i \text{ , and importance} = \gamma_i^2 \text{ with } \sum_{i=1}^{8} \gamma_i^2 = 1 .$$

3 Results

The speed of the steady-state reaction front depends on the bilayer thickness as shown in Figure 4. The thick black line in Figure 4 is the arithmetic average of the 9 LHS samples. The gray area represents the 95% LHS confidence interval of the velocity prediction. The 95% uncertainty in the data varies from ±0.05 to ±0.3 m/s (depending on multilayer geometry) and ±3 nm representing two standard deviations. For example, a bilayer thickness of 10 nm resulted in a predicted front velocity 5-8 m/s. As the bilayer thickness was increased, the front velocity increased to 6-10 m/s at the $t_{b,max}$ thickness of ~15 nm as shown in Figure 4. The front velocity decreased with increasing bilayer thickness when $t_b > t_{b,max}$. Most of the data were within the LHS confidence interval. The velocity of the datum point above 80 nm was less than the model prediction. The mean of the data points near $t_{b.max}$ were also outside of the 95% predicted LHS confidence interval. More work is needed to address these extreme points. Temperature dependent thermophysical properties are probably necessary for better agreement as temperatures range from 300 K to 2000 K. The front may be driven by the ignition source causing more uncertainty for thin bilayer thicknesses near t_{bc}.

Figure 5 shows the parameters that contribute to the uncertainty in the velocity prediction. The percent perturbation about the mean is also given in Figure 5. Since detailed information about the scatter in the eight input parameters is unknown, a simple uniform distribution was assumed. The uncertainty in the predicted velocity is based on the assumed distribution and magnitude of the scatter about each of the eight parameters. The actual scatter may be more or less than the values given in Figure 5. The three parameters that contribute to the uncertainty in the burn velocity prediction are the heat of reaction, diffusion coefficient, and the thermal conductivity of Al. The uncertainty in liquid aluminium is likely higher than the range studied here. For example, the Al thermal conductivity decreases to about 93 W/mK [12] upon melting. Furthermore, the specific heat of Al increases from 897 to 1150 J/kgK as temperature increases from 300 to 800 K [11]. Clearly, temperature dependent properties should be considered for future work.

Figure 6 shows a few of the temperature profiles from 0.05 to 0.5 ms after ignition of a free standing foil with a bilayer thickness of 100 nm.

A constant temperature boundary was used to initiate the Co/Al nanolaminate. The temperature of the left-hand boundary was set to a constant value of 800 K, which was 134 K below the melting point of the aluminium. As the energy was conducted into the foil, the exothermic reaction of Al with Co

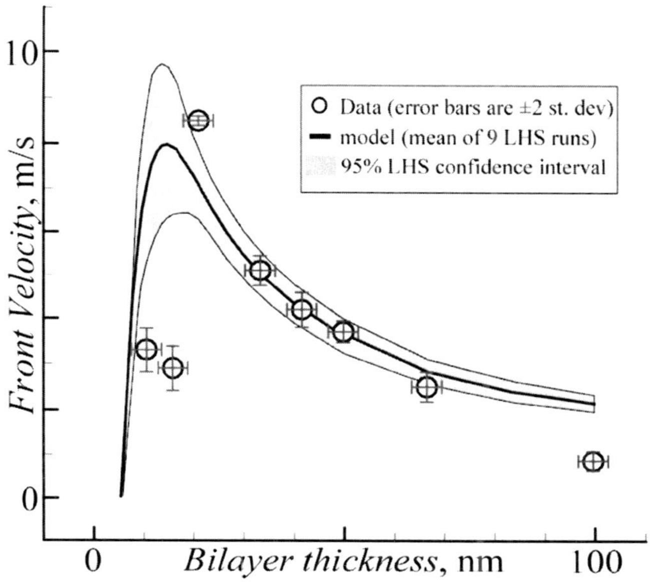

Figure 4: Effect of bilayer thickness on burn front speed. The black line is the mean of 9 LHS simulations. The gray area is the 95% LHS confidence interval. Symbols are measurements with error bars.

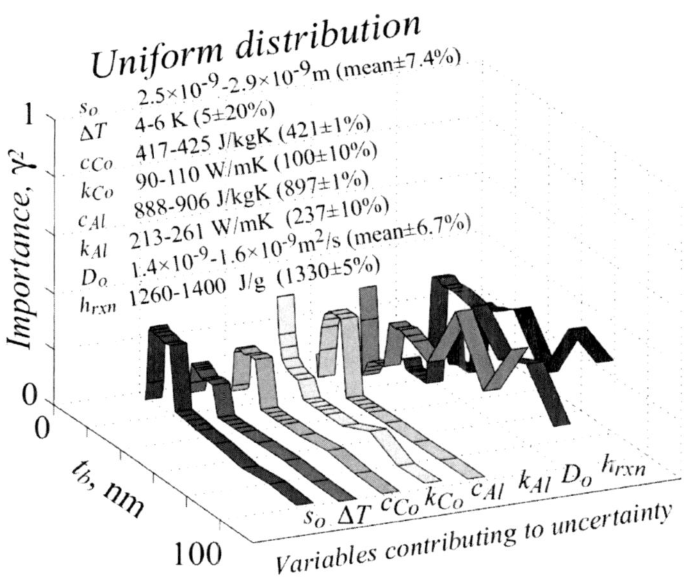

Figure 5: Parameters that contribute to uncertainty in velocity prediction.

caused the temperature to increase above the boundary temperature at 0.05 ms. At 0.2 ms, the exothermic reactions were sufficient to completely melt the aluminium. At 0.27 ms, the temperature rise was sufficient to exceed the melting point of the cobalt. The temperature profiles level out at the melting point of both the aluminium and the cobalt. At steady-state, the temperature rises from 300 K to about 2070 K. The time step used for these "stiff" calculations was 0.1×10^{-9} s.

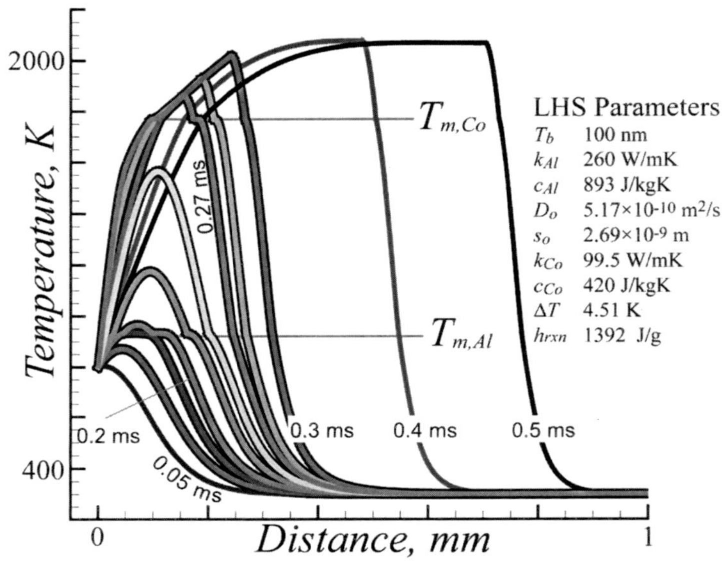

Figure 6: Temperature profiles from 0.05 to 0.5 ms after ignition of a Co/Al nanolaminate with a bilayer thickness of 100 nm.

The front profiles depicted in Figure 6 do not show oscillatory behaviour (i.e. periodic fluctuations in temperature with distance). However, for certain parameter sets, oscillatory behaviour was observed in the temperature profiles. Oscillatory behaviour was most noticeable when the reaction energy was low, such as for thin bilayers. Figure 7 shows temperature oscillations ranging from 1290 to 1294 K for a system with 10.7 nm bilayers. The period of the oscillations was about 100 μm. For a bilayer thickness of 10.7 nm, more than 50% of the bilayer is composed of the premixed Co_xAl_y phase leaving the nanolaminate without sufficient reactants to continue the exothermic temperature increase. Other modes of oscillation related to latent effects have been observed, e.g. the temperature excursions are halted when the melting point is reached. Oscillatory behaviour in exothermic nanolaminates has also been reported by Jayaraman et al. [13].

A rippled surface morphology has been discovered in reacted Co/Al foils (after cool down). Although this could be the result of several different mechanisms, it is intriguing that the structural periodicity is similar to that predicted from modelling. Figure 7 shows a scanning electron micrograph of the surface morphology of a reacted Co/Al foil.

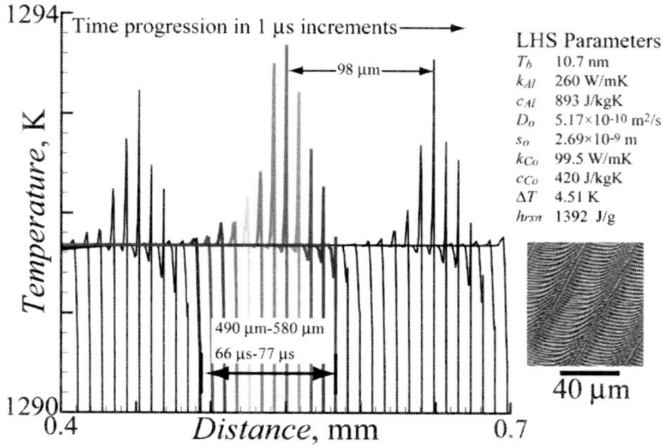

Figure 7: Temperature profiles from 56 to 91 μs after ignition of a Co/Al nano-laminate with a bilayer thickness of 10.7 nm. The scanning electron micrograph (bottom right) shows the post reaction foil surface morphology for a bilayer of 66.4 nm and total thickness of 7.5 μm.

4 Summary and conclusions

The diffusion-limited model of Hardt and Phung [5] has been used with a simple phase-change model to simulate the self-sustained, high temperature front propagation of Co/Al nanolaminates. The model utilizes a stiff ODE solver with adaptive gridding. Both steady and oscillatory front behaviours were predicted. The oscillatory behaviour was most prominent for small bilayer thickness where more than half of the bilayer was composed of a premixed Co_xAl_y layer. This finding is consistent with predictions by Jayaraman et al. [13] for Ni/Al nanolaminates. The dependence of the front velocity on the bilayer thickness has been predicted and compared to experiments. The uncertainty of the front velocity to various model parameters was also determined with the most sensitive parameters being the reaction enthalpy, diffusion coefficient, and thermal conductivity of the Al. The measured front velocities for a large range of bilayer thicknesses up to 80 nm are within the 95% LHS prediction interval. However, the model indicated that the front velocities for bilayer thicknesses above 80 nm were faster than measured front velocities. Differences may be due to using constant thermophysical properties for Co/Al. Future work will address temperature dependent properties and other foil compositions.

Acknowledgements

Sandia is a multiprogram laboratory operated by Sandia Corporation, a Lockheed Martin Company, for the United States Department of Energy under Contract DE-AC04-94AL85000. The authors appreciate the technical assistance of Eric

Jones, Jr. We are thankful to Mel Baer for suggesting the use of the Hardt and Phung [5] diffusion model. We are also thankful for comments from internal reviewers at Sandia National Laboratories, Rob Sorensen and Bill Erikson.

References

[1] Wang, J., Besnoin, E., Knio, O.M., Weihs, T.P., Effects of physical components on reactive nanolayer joining. *Journal of Applied Physics*, **97**, pp. 114307(1)-7114307(7), 2005.

[2] Adams, D.P., Hodges, V.C., Bai, M.M., Jones, E. Jr., Rodriguez, M.A., Buchheit, T., and Moore, J.J., Exothermic Co/Al nanolaminates. *Journal of Applied Physics*, in review.

[3] Wickersham, C.E., and Poole, J.E., Explosive crystallization in zirconium silicon multilayers. *J. Vac. Sci. Technol. A, **6(3)**, pp. 1699–1702, 1988.*

[4] Hobbs, M.L. and Adams, D.P., Reactive nano-films of Al and Pt. *Proceedings of the 8th World Congress on Computational Mechanics*, Venice, Italy, June 30-July 5, 2008.

[5] Hardt, A.P. and Phung, P.V., Propagation of gasless reactions in solids—I. Analytical study of exothermic intermetallic reaction rates. *Combustion and Flame*, **21**, pp. 77–89, 1973.

[6] Gross, R.J., Baer, M.R., and Hobbs, M.L., XCHEM-1D: a heat transfer-chemical kinetic computer program for multilayered reactive materials. Sandia National Laboratories Report SAND93-1603, 1993.

[7] Yao, L.S. and Prusa, J., Melting and freezing. *Advances in Heat Transfer*, **19**, Hartnett JP and Irvine TR Jr., editors (1989).

[8] Lide, D.R., editor-in-chief, CRC Handbook of Chemistry and Physics 88th Edition 2007-2008, CRC Press, Cleveland, Ohio, 2008.

[9] de Boer, F.R., Boom, R., Mattens, W.C.M., Miedema, A.R., and Niessen, A.K., Cohesion in Metals Transition Metal Alloys, Elsevier Science Publishers, North-Holland Physics Publishing, Amsterdam, 1988.

[10] McKay, M.D., Conover, W.J., and Beckman, R.J. A comparison of three methods for selection values of input variables in the analysis of output from a computer code. *Technometrics*, **22(2)**, p. 239, 1979.

[11] Incropera, F.P. and DeWitt, D.P., *Fundamentals of Heat and Mass Transfer*, 5th Edition, John Wiley & Sons, New York, 2002.

[12] Weast, R.C., Editor. CRC Handbook of Chemistry and Physics 55th Edition. CRC Press, Cleveland, Ohio, p. E10, 1974.

[13] Jayaraman, S., Knio, O.M., Mann, A.B., and Weihs, T.P., Numerical predictions of oscillatory combustion in reactive multilayers. *Journal of Applied Physics*, **86(2)**, pp. 800–809, 1999.

Heat transfer and pressure drop experimentation inside single minichannels

A. Cavallini, S. Bortolin, D. Del Col, M. Matkovic & L. Rossetto
Dipartimento di Fisica Tecnica, University of Padova, Italy

Abstract

The present authors have designed and built a test apparatus for heat transfer and pressure drop measurements during forced convective single phase flow, condensation and flow boiling in one single minichannel. The performance of the test apparatus for investigation of heat transfer and fluid flow is described in this paper. Some construction issues with important effects on the experimental technique are also reported in the paper. The adiabatic and diabatic two phase pressure drop, the local heat transfer coefficient and the dry-out phenomenon, as well as the single phase heat transfer coefficient measurements, are all discussed herein. The experimental apparatus has turned out to be highly effective in terms of low experimental uncertainty, local measurement capabilities and rather fast data acquisition rate over a wide range of test conditions. Vast, reliable experimental data have a key role in semi-empirical modelling. Indeed, these experimental results, together with data associated to minichannels from different authors, have been used for the development of new heat transfer and pressure drop models able to accurately predict experimental data for low, medium and high pressure fluids.

Keywords: heat transfer, pressure drop, minichannels, condensation, flow boiling, dry-out.

1 Introduction

In the field of heat transfer mini-, micro- and nanochannels have gained a great deal of attention in engineered systems. A number of advantages associated with minichannels, with an inner diameter in the range of 0.2-3 mm [1], have pushed their use far ahead of the actual understanding of the microscale process and how the scaling down alters the governing physical processes. First compact heat

WIT Transactions on Engineering Sciences, Vol 61, © 2008 WIT Press
www.witpress.com, ISSN 1743-3533 (on-line)
doi:10.2495/HT080131

exchangers in the automotive industry, for example, were integrated into vehicles' air-conditioning systems before they were really optimized in terms of fluid charge, material use and compactness. Optimizing the device performances reduces its size, energy requirements, costs and last, but not the least, impact on the environment [2].

Minichannels adopted in heat transfer engineering can be found in a wide variety of applications: from residential air-conditioning to spacecraft thermal control. Compact elements work with reduced refrigerant charges and can usually withstand extremely high system pressures. A small charge reduces the direct greenhouse effect of the systems and promotes the use of flammable or even toxic refrigerants. Furthermore, resistance to high pressure enables one to use natural refrigerants, such as carbon dioxide, in these systems and hence the operation in the supercritical region.

In order to better understand how the scaling-down alters fluid flow, heat transfer and phase change processes, the present authors designed a special test apparatus for investigation within single minichannels (Matkovic [3]).

2 Short overview of the test apparatus

With reference to the test rig in fig. 1, subcooled refrigerant from the postcondenser passes the mechanical filter and dehumidifier. It is pumped with a variable speed electric motor gear pump through the Coriolis effect mass flow meter and the throttling valve into either the evaporator or the sub-cooler, depending upon the desired test condition at the inlet of the test section. Here, the temperature and the pressure measurements define the thermodynamic state of the refrigerant.

An important part of the primary loop is the accumulator (PV). It is used to store a certain amount of the test fluid as a liquid. The system pressure in the rig is controlled by varying the volume of the liquid contained in the accumulator. Since the liquid contained in the accumulator after steady state operation has been reached is the redundant amount of the system fluid, its size has been carefully defined. As the accumulator is partly filled with a compressible gas, such as nitrogen, it works also as an expansion vessel. Therefore, this element should never be detached from the system.

Another important element of the primary refrigerant loop is the throttling valve (TV2). It is installed after the pump, accumulator and Corriolis effect mass flow meter. Its purpose is to make the fluid flow more stable with proper throttling.

The "reset" bypass (dashed line) allows one to send some fluid back to the inlet of the postcondenser. In this way, the pump works at rather high flow rates, and the part of the test rig from the postcondenser down to the outlet of the pump is sufficiently subcooled to avoid possible cavitation. As a consequence, the flow is more stable. The cavitation is likely to occur at low flow rates for the small thermal capacity, and could cause insufficient feeding of the pump and lack of fluid flow through the test section. It is therefore important to have sufficient reflux of the refrigerant, particularly when the saturation temperature is relatively close to the ambient temperature.

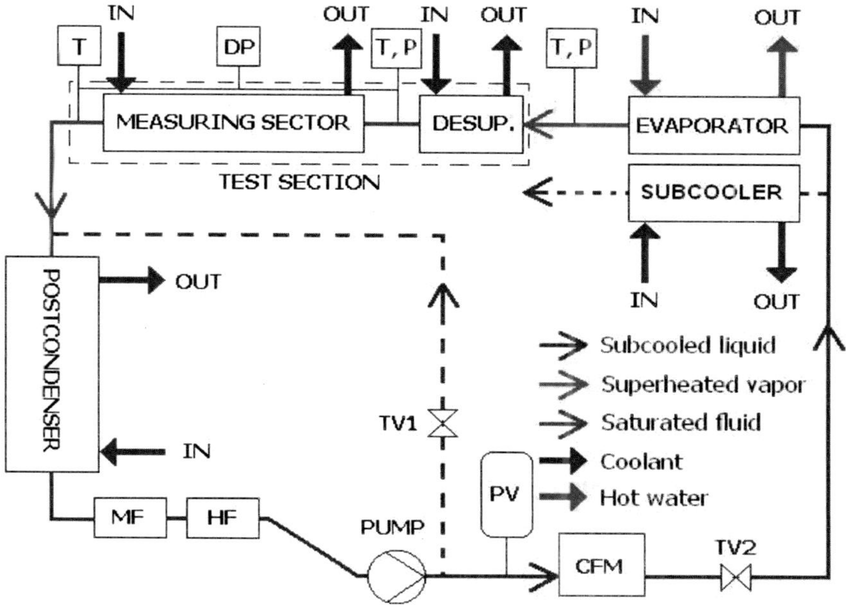

Figure 1: Scheme of the test rig.

The most important part of the test rig is the test section, which consists of two sectors. The first one, the pre-sector (PS) is used to prepare the desired inlet vapour quality, whereas the second one is actually the measuring sector (MS). Between the two sectors and downstream the MS, adiabatic stainless steel pipes are installed in order to reduce the axial heat conduction through the tube wall, to thermally detach the two sectors from the surrounding and to check the saturation state of the fluid by measuring the adiabatic wall temperature and the fluid pressure.

The test section is equipped with a number of thermocouples both in the wall and in the coolant channel along the sector; it is made from an 8 mm copper rod with a 0.96 mm internal bore – the minichannel.

Figure 2: Close-up of the measuring sector during construction.

A rather tortuous path for the secondary fluid is machined in the thick copper wall surrounded by an epoxy resin sheath (fig. 2) working as an insulator, armature and support for thermocouple wires. Crucial advantages of such a design reflect in the following characteristics:

- good coolant mixing and thus accurate temperature measurements along the measuring sector is essential to obtain a reliable water temperature profile, which is used for determination of local heat flux;
- both the continuous interrupting of the boundary layer, due to the rather complex coolant flow passage geometry, and the enhanced external wall surface area notably decrease the external heat transfer resistance, which is crucial for precise heat transfer coefficient measurements;
- thermocouple wires embedded in the copper wall are led outside the measuring sites through the epoxy resin without passing through the coolant flow: in this way the error of the temperature measurements due to axial heat conduction along the thermocouple wire and the spurious emf's build up for the presence of high temperature gradients is reduced to a minimum;
- The epoxy resin sheath does not only serve for accommodation of the thermocouple wires and for the insulation purpose; it also plays an important role as the test section's support.

2.1.1 The test section length

One of the first steps required for the test section design was the calculation of the test tube length. The coolant flow rate in the measuring sector that was initially built for the condensation studies was obtained from the calculated condensation heat flow rate and the desired overall coolant temperature rise in the sector. It was shown that, for a given test condition, any deviation of the tube length from the optimum one increases the experimental uncertainty of the test facility. Since the test tube was considered as a condenser, its size (length) depends severely on the external-side single-phase heat transfer. Good characteristics of the minichannel test tube require:

- High external heat transfer coefficient.
- Enhanced external surface area.
- Homogeneous wall surface temperature distribution along the channel.
- Low thermal resistance of the channel wall in radial direction.
- High thermal resistance of the channel wall in axial direction.
- Good coolant mixing.
- Low pressure drop.

The first two characteristics enhance the external heat transfer, which moves the leading thermal resistance toward the refrigerant side. In this way, the wall to refrigerant temperature difference is increased at given heat flow rate, while the relative error of the corresponding temperature difference measurement is decreased. Enhanced external heat transfer should be achieved so as to avoid systematic errors in wall temperature measurement due to local temperature variations. Furthermore, a high thermal conductivity of the tube wall decreases the associated temperature gradients and thus the wall temperature error due to deviation in temperature sensor positioning. On the other hand, the high thermal

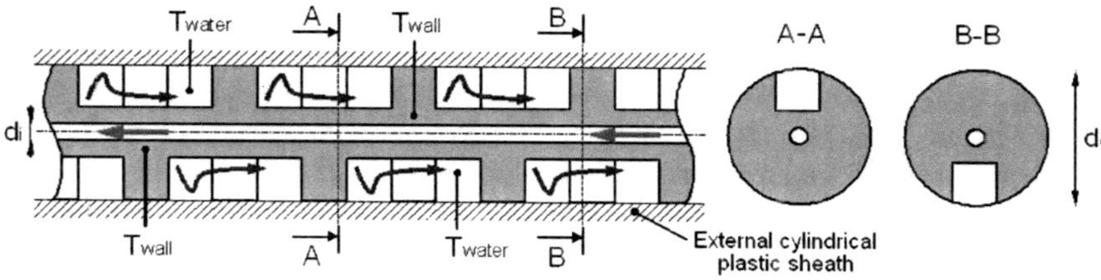

Figure 3: Schematic of the coolant flow passage geometry.

conductivity of the test tube also promotes the axial heat conduction. Even though, much smaller in comparison with the radial one, the axial heat conduction is additionally reduced by the multiple groves in the wall thickness of the present design. Additionally, a precise coolant temperature profile measurement is also of paramount importance for high performance heat transfer coefficient measurements. Insufficient coolant mixing is probably one of the most frequent reasons for systematic errors in fluid temperature measurement. On the other hand, a flow passage geometry that enables good fluid mixing and enhanced heat transfer are usually associated with significant pressure drop. In this context, the present coolant flow passage geometry has turned out to offer an excellent performance behaviour for high precision HTC measurements inside a single minichannel.

2.1.2 Sensitivity of the temperature profile measurements

Once the test section length and the test conditions were decided, positions for wall and coolant thermocouples had to be established. Optimal spacing between the temperature sensors depends on their nominal uncertainty and the temperature gradients along the test section. A minimum distance between two neighbouring thermocouples was defined by considering the axial displacement required for a minimum temperature rise exceeding the nominal error of the elements.

The minimum rational distance between two neighbouring thermocouples is defined according to the local temperature gradient along the measuring sector. In fact, a high heat transfer rate at the test section inlet results in high water and wall temperature gradients and thus a shorter distance between two thermocouples is needed to exceed the uncertainty of thermocouples. During condensation along the test tube, lower gas velocity induces lower heat transfer rate that requires less thermocouples to be set. The actual number of installed temperature sensors comes up as a compromise between a good temperature profile description and a reasonable design. Finally, there are 31 equally spaced thermocouples installed in the measuring sector: 16 in the wall and 15 in the water channel, both measuring reliable temperature profiles.

2.1.3 Correction to the wall temperature measurement

To estimate the correction of the wall temperature measurement due to the thermal resistance between the thermocouple's position and the inner wall surface, the temperature profiles around the central bore of the test tube have been plotted in fig. 4. When the thermocouples are positioned within a certain distance from the inner wall surface, a constant shifting of the measured wall temperature dependent on the heat flux is observed despite the variable external test conditions. For a thermocouple positioned 0.7 mm from the inner wall surface, temperature readings would result in around 0.23 °C lower values, for a given boundary condition, as compared to the internal wall surface temperature. Indeed, with respect to the acquired test conditions, appropriate correction for the wall temperature measurement is considered during data reduction.

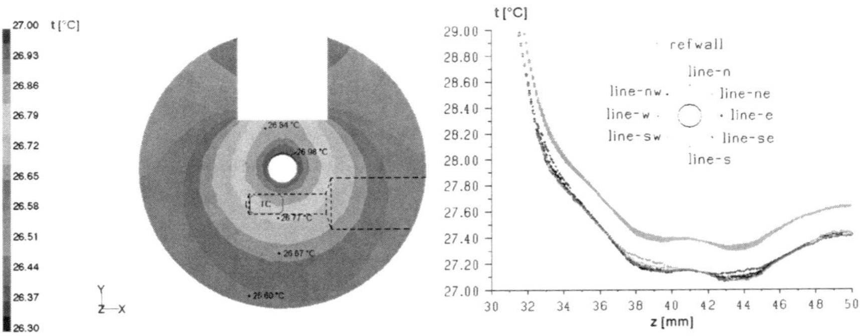

Figure 4: Wall surface temperature distribution.

3 Experimental measurements

3.1 HTC during single phase flow, forced convective condensation, flow boiling and post dry-out

Heat transfer coefficient measurements have been performed during single phase vapour and liquid flow, forced convective condensation, flow boiling and dry-out (figs. 5, 6). Poor resistance to high temperatures of the epoxy resin (<80 °C), maximum system pressure of the gear pump head (<30 bar) and freezing point of the coolant – water (0 °C) are the main system constraints. Schematics of the studied phenomena with R134a flowing within the single minichannel are shown in log p-h diagram of the figure 5. Single phase heat transfer coefficient measurements during liquid and vapour flow have been performed for both cooling and heating modes. In single phase flow good agreement has been found of average experimental values versus the Gnielinski correlation [4, 5].

Besides, a precise measurement of the condensation heat transfer coefficients within the single minichannel has also been carried out (fig. 7 left). The present technique permits local measurements of the heat flux and saturation to wall temperature difference. This data is rare in literature since most authors measured condensation HTCs within multiport minichannels [6].

The averaged values of the heat transfer coefficient over the number of parallel minichannels do not give sufficient information on the phenomenon within a single minichannel. Besides, most of the experiments have been

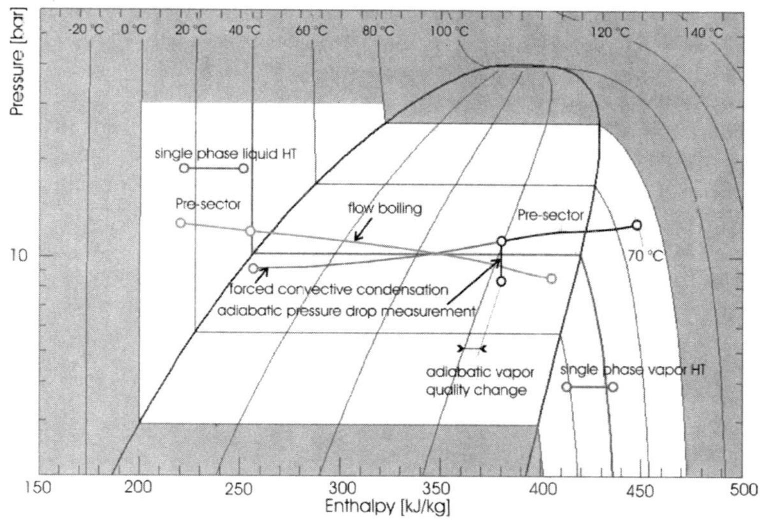

Figure 5: Different heat transfer coefficient and pressure drop measurements in log p-h diagram for R134a.

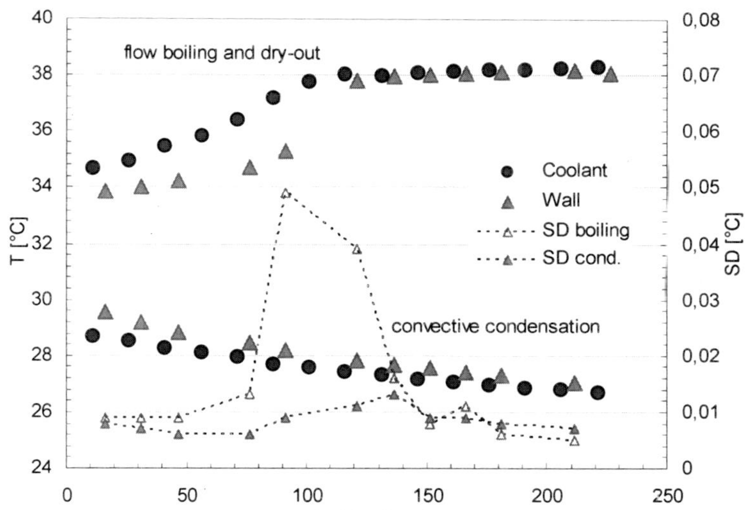

Figure 6: Temperature profiles for flow boiling and forced convective condensation HTC experiments with R134a at G=200 kg m^{-2} s^{-1}.

performed with the Wilson plot technique, which does not require the wall temperature measurement for the heat transfer coefficient deduction, but calculates the coefficients from the variation of the external thermal resistance.

This technique overpasses the difficulties raised by a reliable wall temperature measurement, but on the other hand, it presents a significant experimental error when the leading thermal resistance is on the coolant side.

Furthermore, local flow boiling heat transfer coefficients have also been measured with the present test apparatus (fig. 7 right). Literally the same measuring technique as the one adopted for condensation studies has been used [7, 8]. In addition to temperature profiles, local temperature standard deviation values have turned out to be of crucial help for the definition of the dry-out region. There, during dry-out, the liquid film at the wall dries up with an oscillating process, leading to alternate change in thermal resistance. The region with rapid change in thermal resistance is thus denoted by a significantly higher wall temperature standard deviation as compared to the adjacent regions subject to boiling process or forced convection with gas (fig. 6).

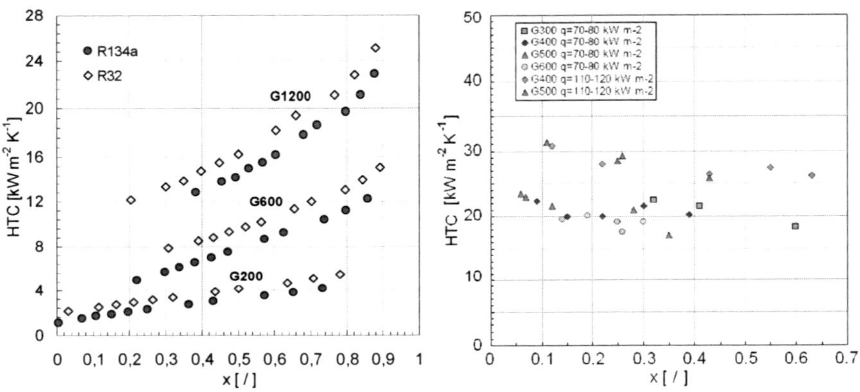

Figure 7: Experimental heat transfer coefficients versus vapour quality during condensation of R134a and R32 (left) and during flow boiling of R134a (right) inside a single minichannel.

3.2 Pressure drop measurements during single phase flow, adiabatic and diabatic vapour-liquid flow

Pressure drop experiments have been performed with two different fluids over a wide range of test conditions: during single-phase liquid and vapour flow (fig. 8 left), adiabatic two-phase flow (fig. 8 right) and during condensation flow. The experimental friction factor has been calculated from the overall measured pressure drop, the calculated values for two abrupt cross section area changes and the friction pressure drop within the adiabatic stainless steel capillaries [9], all included within the pressure measuring section. The experimental single phase friction factor is in good agreement with the Churchill [10] equation, which is reported over the entire span of Reynolds number. With regard to diabatic tests, pressure recovery due to condensation, boundary conditions for the two sudden geometry changes and mean value for the two phase multipliers

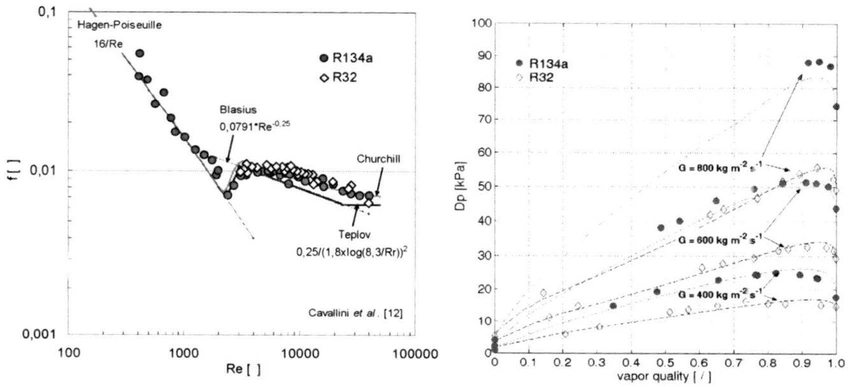

Figure 8: Experimental single phase friction factor versus Re (left) and overall two phase pressure drop vs. vapour quality (right).

have all been accounted for in the data reduction. The two-phase multiplier was calculated following Cavallini et al. [9] while the local pressure losses at the abrupt geometry changes were modelled according to Paliwoda [11].

4 Conclusions

The performance of the test apparatus for heat transfer and fluid flow studies within a single minichannel is described through the discussion of some design aspects and presentation of experimental test runs. This apparatus and the experimental technique for local HTC measurements with the use of a secondary cooling-heating medium have turned out to be highly efficient for experimental studies within a single minichannel.

References

[1] Kandlikar S.G. and Grande W.J., Evolution of microchannel flow passages – thermohydraulic performance and fabrication technology, Heat Transfer Engineering 24(1) (2003) 3–17.

[2] Kandlikar S. G., Garimella S., Li D., Colin S., King M. R., Heat transfer and fluid flow in minichannels and microchannels, Elsevier Ltd., ISBN: 0-0804-4527-6, 2006.

[3] Matkovic M., Experimental condensation inside minichannels, Ph.D. Thesis, Università di Padova, 2006, 209 p.

[4] VDI heat atlas / ed. Verein Deutscher Ingenieure, VDI-Gesellschaft Verfahrenstechnik und Chemieingenieurwesen (GVC), [Transl.: J. W. Fullarton]. – Düsseldorf: VDI-Verl., 1993.

[5] VDI-Wärmeatlas, Verein Deutscher Ingenieure, VDI-Gesellschaft Verfahrenstechnik und Chemieingenieurwesen (GVC), Springer-Verlag Berlin Hedelberg New York, 2002

[6] Cavallini, A., Doretti, L., Matkovic, M. and Rossetto, L., Update on condensation heat transfer and pressure drop in minichannels, Heat Transfer Engineering 27(4) (2006a) 74–87.

[7] Cavallini A., Bortolin S., Del Col D., Matkovic M., Rossetto L., Local Heat Transfer Coefficient During Flow Boiling inside Circular Minichannel, ASME-JSME Thermal Engineering Summer Heat Transfer Conference, July 8–12, 2007, Vancouver, British Columbia, Canada.

[8] Del Col D., Cavallini A., Bortolin S., Matkovic M., Rossetto L., Heat Transfer Coefficient during Flow Boiling of R134a in a Circular Minichannel, 5th European Thermal-Sciences Conference, 18–22 May, 2008, Eindhoven, Netherlands.

[9] Cavallini A., Del Col D., Matkovic M., Rossetto L., Frictional Pressure Drops During Vapour-Liquid Flow in Minichannels: Experimental Data and Modelling, Micro/Nanoscale Heat Transfer International Conference, Tainan, Taiwan, 2008, MNHT2008-52282.

[10] Churchill, S.W., "Friction factor equation spans all fluid-flow regimes", Chemical Engineering 45, 1977, pp. 91–92.

[11] Paliwoda, A., "Generalized method of pressure drop calculation across pipe components containing two-phase flow of refrigerants", Rev. Int. Froid, 15(2), 1992, pp. 119–125.

[12] Cavallini A., Del Col D., Matkovic M., Rossetto L., Pressure Drop During Two-Phase Flow of R134a And R32 in a Single Minichannel, Submitted to Journal of Heat Transfer, 2008.

Section 6
Modelling and experiments

Advances in gas turbine blade cooling technology

R. S. Amano
University of Wisconsin-Milwaukee, USA

Abstract

This article presents a background of the gas turbine blade cooling technologies along with numerical methodologies and physical models that are most commonly used in the computations of blade flows in gas turbine blades. In addition some advancement in the cooling technologies is also discussed.

Keywords: high temperature flow field, gas turbine, advanced cooling technology.

1 Introduction

To increase the efficiency and the power of modern power plant gas turbines, designers are continually trying to raise the maximum turbine inlet temperature. Over the last decade the temperature has risen from 1500 K to 1750 K in some high-performance units. With this increase of the temperature only about 25% can be attributed to improved alloys. New materials, such as ceramics, could help increase this maximum temperature even more in the future. However, most of the recent improvements in inlet temperature come from better cooling of the blades and a greater understanding of the heat transfer and the three-dimensional temperature distribution in the turbine passage. Higher gas temperature generally causes increased blade temperature and greater temperature gradients, both of which can have a detrimental effect on service life. As of today improvements in computational techniques in turbomachines have been attempted by industrial researchers because the numerical approaches are quite advantageous in comparison with experimentation, due to its ease of modeling, relatively complicated geometry and an unsteady flow nature. In fact, modern turbomachinery operates under extremely complex three-dimensional flow conditions, and further improvement in performance requires detailed

www.witpress.com, ISSN 1743-3533 (on-line)
doi:10.2495/HT080141

knowledge of the flow structure. Particularly, the need to estimate operating conditions, secondary flows, and turbulence and heat transfer rates demands that viscous models be examined. Near the hub and tip of a turbine stator/rotor passage, the flow is affected by the interaction between the stream-wise boundary layer and the side-wall boundary layer. Although this region is thin, its effect on the overall aerodynamic performance cannot be neglected. To design a high performance turbine, an engineer has to understand the detailed three-dimensional flow field near the hub and tip. The flow near the midspan of a turbine stator passage is, principally, driven by the inviscid process. However, some important characteristics and flow parameters are strongly influenced by the turbulence transport near the solid walls and the wake region behind the airfoils. Under certain operating conditions, the boundary layer development on the blade surface is much enhanced due to the existence of adverse pressure gradients, which have considerable effect on the following rotor stage. For the reasons mentioned above, a technique for performing three-dimensional computations becomes more and more necessary. In the past decade computational fluid dynamics (CFD) has undergone significant evolution in turbomachinery research. Coincident with the substantial enhancements in computer technologies, three-dimensional computations became more and more useful.

This paper summarizes the most advanced cooling technologies that are currently used in USA and EUROPE for the power generation systems as well as aerospace projects.

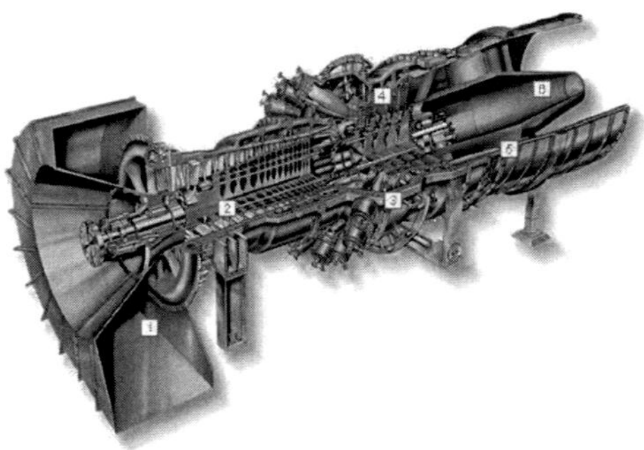

Figure 1: Combustor and high-pressure turbine.

Cooling technology, as applied to gas turbine components is composed of five main elements, (1) internal convective cooling, (2) external surface film cooling, (3) materials selection, (4) thermal-mechanical design, and (5) selection and/or conditioning of the coolant fluid. Cooled turbine components are merely highly specialized and complex heat exchangers that release the cold side fluid in a controlled fashion to maximize work extraction. The enhancement of internal convective flow surfaces for the augmentation of heat transfer was initially improved some 25 to 30 years ago through the introduction of rib-rougheners or turbulators, and also pin-banks or pin-fins. Figure 2 shows an example schematic of a blade cooling circuit that utilizes many turbulated passages, a pin bank in the trailing edge, and impingement in the leading edge (coolant is released via film holes, tip holes, and trailing edge). These surface enhancement methods continue to play a large role in today's turbine cooling designs. Film cooling is the practice of bleeding internal cooling flows onto the exterior skin of the components to provide a heat flux reducing cooling layer, as shown by the many holes placed over the airfoil in Figure 2. Film cooling is intimately tied to the internal cooling technique used in that the local internal flow details will influence the flow characteristics of the film jets injected on the surface.

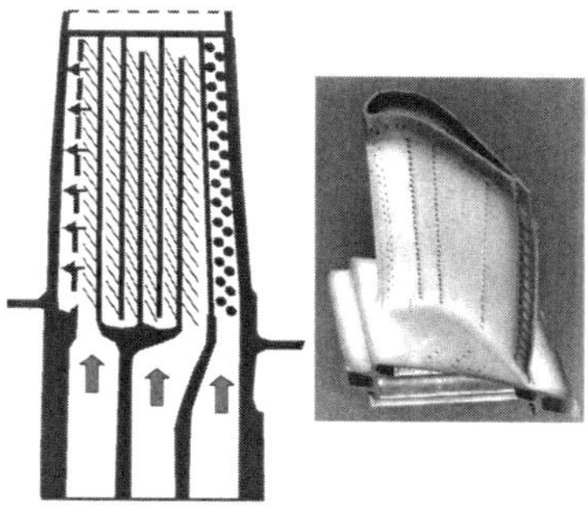

Figure 2: Example cooled turbine blade and cooling circuit.

Several characteristics of gas turbine cooling are worthy of note prior to describing any specific technologies. Almost all highly cooled regions of the high-pressure turbine components involve the use of turbulent convective flows and heat transfer. Very few if any cooling flows within the primary hot section are laminar or transitional. Moreover, the typical range of Reynolds numbers for cooling techniques, using traditional characteristic lengths and velocities, is from 10,000 to 60,000. This is true of both stationary and rotating components. The enhancement of heat transfer coefficients for turbine cooling makes full use of the turbulent flow nature by seeking to generate mixing mechanisms in the coolant flows that actively exchange cooler fluid for the heated fluid near the walls. These mechanisms include shear layers, boundary layer disruption, and vortex generation. In a marked difference from conventional heat exchangers, most turbine cooling means do not rely on an increase in cooling surface area, since the available surface area to volume ratios are very small. Surface area increases are beneficial, but are not the primary objective of enhancements. The use of various enhancement techniques typically results in at least 50% and as much as 300% increase in local heat transfer coefficients over that associated with fully developed turbulent flow in a smooth duct.

Numbers of researchers have been attaching the problem of improving the cooling technologies on these issues. As a demonstration, the author is presenting recent developed computational studies in the blade cooling computations in the following sections.

2 Internal cooling

Three turbulence models, *k-ε*, Linear *k-ω*, and RSM were used to predict the streamwise mean velocity (m/s) distributions. Figure 3 shows streamwise mean velocity contours at Mid-Plane for Re=36,000, where the flow separation occurs at 45° in the curvature of the bend. All models show some vortex formation at the corners. This is due to the geometry. The flow separation starts at around 45°, the location of the reattachment points change, for RSM this point is around 24 times the curvature diameter. For *k-ε*, the point is about 14 times the curvature diameter. For k-ω, this model shows that there are two vortexes; one occurs about 10 times the curvature diameter and second occurs about 20 times the diameter. The secondary flow gradually disappears around 22 times the curvature diameter

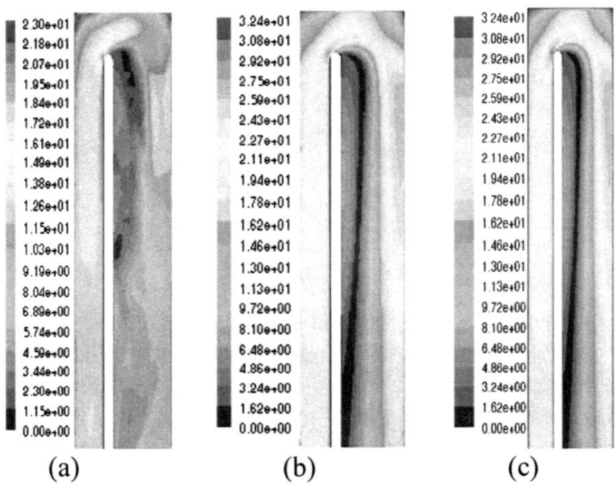

Figure 3: Mean flow along the mid-plane (m/s) (a) RSM (b) *k-ε* and (c) *k-ω* at Re=36000.

Depending on the Reynolds number, the flow strength changes causing different separation zones. The flow after the bend is not free to move but is bounded by the outer wall. This causes the fluid to bounce off the wall. This occurs sooner for higher Reynolds number.

All three models predict nearly a similar trend of the Nusselt number, higher in the bend and reducing gradually to downstream. The Nusselt number development in the upstream section is similar and there is an enhancement in Nusselt number within and after the bend, reaching its highest value. This implies that, as is well established in others cases, within the bend, the flow separation, flow impingement and strong cross duct motion, not only cause a reduction in the thickness of the viscous sub-layer, but also reduce its sensitivity to Reynolds numbers. At the downstream section the normalized Nusselt number decreases because the secondary flows gradually reduce, also due to increase in cross section, which causes a decrease in the Reynolds number.

The k-ω model is poor in capturing high turbulence flow as can be observed. Figure 4 shows the mid-plane mean velocity distribution in the vector form, for experimental [1] and the three models at Re=36,000. The agreement between the experiment and the computation is well, in particular quite matches with the computations using RSM. The curvature-induced separation bubble along the inner wall is considerably smaller than the observed by Cheah et al. [3]. Because of the present geometry, the effective cross sectional area reduces over the second half of the bend, imposing an overall acceleration of the flow. However, additional separation bubbles are formed in the corners.

Figure 5 shows the contour normalized Nusselt number for the Experiment [2] and for the three models at Re= 36,000. These comparisons are also done in

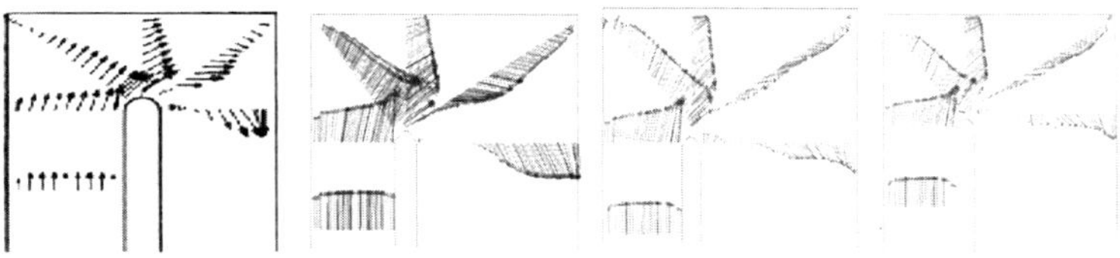

(a) (b) (c) (d)

Figure 4: Mean flow along the mid-plane (m/s) Exp. (b) k-ε (c) k-ω (d) RSM at Re=36,000.

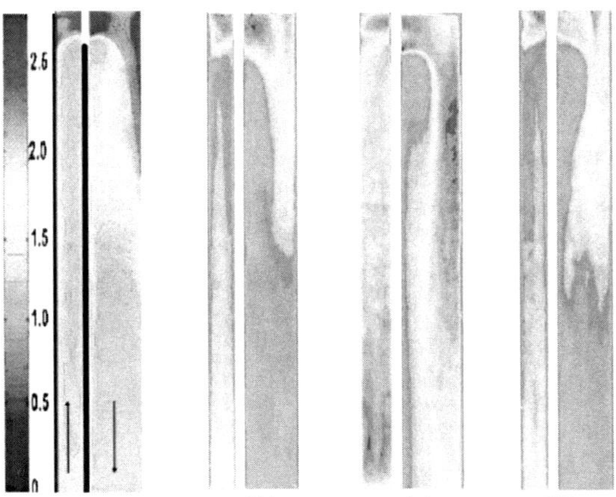

(a) (b) (c) (d)

Figure 5: Nusselt number contour, normalized by the Dittus-Boelter correlations for Nu (a) Exp (b) k-ω (c) k-ε (d) RSM at Re=36,000.

Fig. 6, which shows the normalized side averaged Nusselt numbers (Nu_{avg}/Nu_o) plots at Re=36,000. Here the negative X/D refers to upstream and positive refers to downstream. At X/D=0 the wall is insulated that has high influence on the CFD analysis. This causes deviation observed. As shown in the figures, the agreement is quite good with RSM and also fairly well with *k-ω* model. However, the computations with *k-ε* model show relatively poor agreement with the experiments.

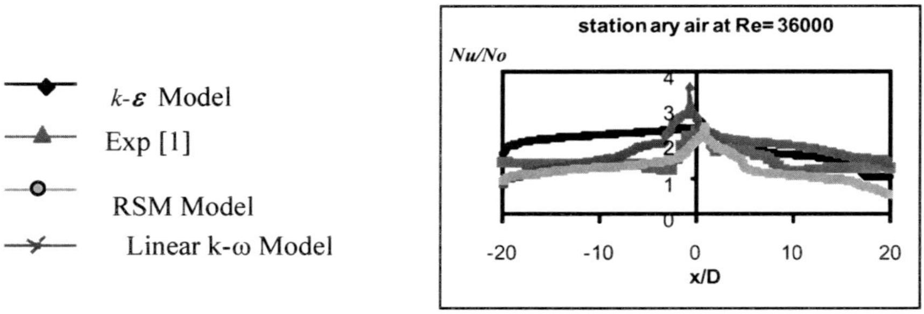

Figure 6: Side-averaged Nu. number for air (Pr=0.71).

3 Current trend in cooling technologies

Figure 7 shows the Nusselt number distribution along the ribbed channels for chevron inline, chevron strips staggered and simple 45 degree angled ribs compared with the smooth channel. As shown in the figure the cases with

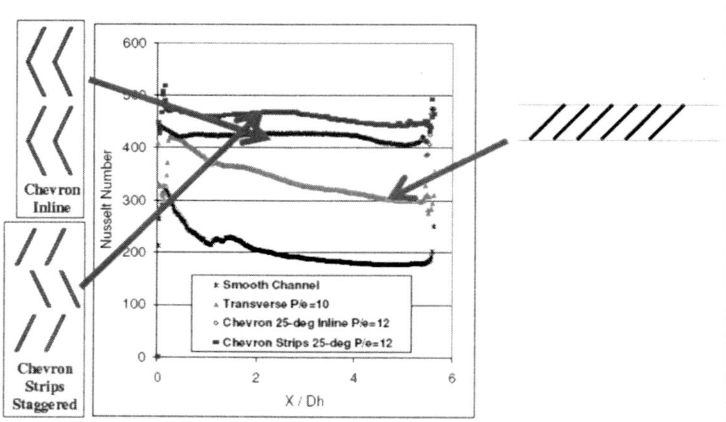

Figure 7: Ribbed channel performance.

chevron arrangements are in general more effective in keeping the Nusselt number at a higher level throughout the channel.

Full cooling distribution is obtained, and a form of transpiration-like film cooling results from the normal holes in the outer layer. Examples of the overall cooling effectiveness and the film cooling are shown in Fig. 8 by Nakamata et al. [4]. The challenges of this micro cooling include hole plugging, wall strength, film cooling, manufacturing, and cost. No commercial use of these micro cooled solutions has yet appeared.

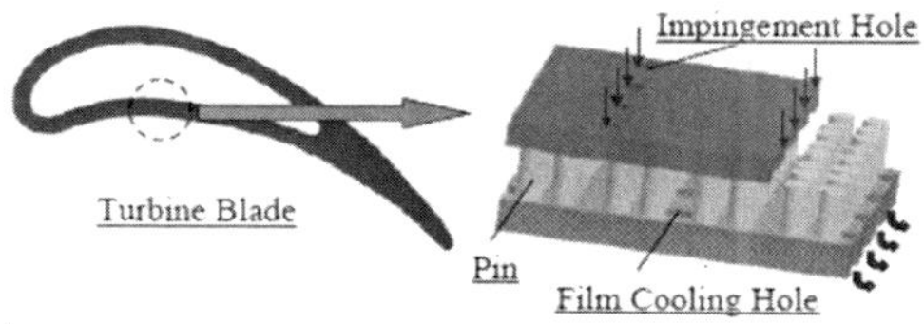

Figure 8: Micro cooling of gas turbine blade [4].

The ultimate goals in gas turbine thermal management include (1) isothermal components to eliminate all internal thermal stresses, (2) hot components to minimize heat loads and inefficient energy transfers in the engine, and (3) minimization of thermal gradients between components and its structures. The advancement of turbine cooling has allowed engine design to exceed normal material temperature limits, but it has shown complexities that have accentuated the heat transfer cooling performances greatly. Cooled component design has consistently trended in the direction of higher heat loads, higher through-wall thermal gradients, and higher in-plane thermal gradients. Gas turbine heat transfer and thermal management technology advancements over the last two decades have primarily come in the more detailed understanding of boundary conditions and the higher sophistication of analysis tools.

Now there are several issues to be considered in further advancing the gas turbine cooling technology. Some technologies such as internal blade cooling and film cooling challenges address active cooling technologies, hot gas flow paths, component design, and systems design. There may not be existing specific solutions, or to prioritize the issues yet as far as the ones to push forward. However, in order to demonstrate some means of comparison, Fig. 9 depicts an approximate ranking of the issues on the basis of technology impact versus technology risk. Such risk-benefit rankings are commonly used to determine a balanced portfolio of funded technology projects within businesses, including research and development. Technology impact includes engine performance metrics and range of product applications, while technology risk refers to the probability of successful development and deployment. The region near the upper-left represents a technology accounting for immediate development; whereas the region near lower-right corner represents the very low benefit-to-risk ratio items that should be avoided. All of these technologies discussed here are seen to cover a broad spectrum of impact and risk, though

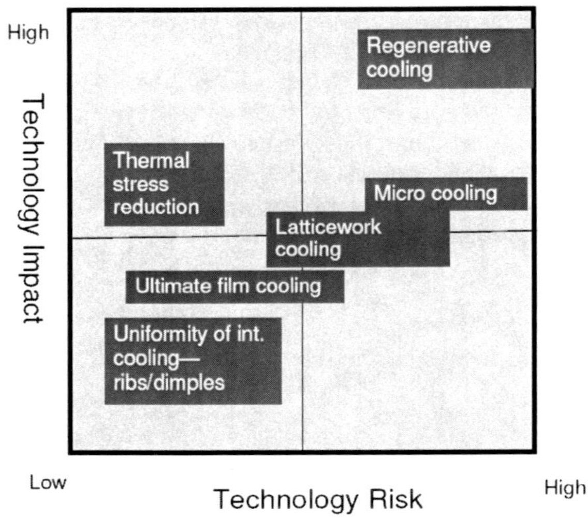

Figure 9: Technology risk vs. technology impact.

individual opinions on rankings may differ. Those items on the left half of the chart mainly represent the issues most strongly being addressed by research today. Those higher risk items on the right side of the chart represent potentially fruitful directions requiring significant innovations and research. Singling out the highest risk-benefit technology identified in Fig. 8, regenerative cooling may cover extensive applications and formats from fairly simple concepts on components to very complex system plant designs. While some versions of regenerative cooling may not entail this high risk, such as those in practice today, this ranking is used to indicate the higher degree of total cycle complexity and risk for most broadly based concepts.

4 Conclusions

This summary has presented several of the cutting edges, innovative cooling methods expected to further enhance the aero-thermal-mechanical performance of turbine engines used for power systems. These methods are by no means an exhaustive or comprehensive summary however. Many other variations and combinations of these techniques are anticipated as manufacturing advances become reality. Further improvements and new techniques may become feasible as materials, systems integration, and controls also advance.

Research trends and potential solutions to these thermal issues have been briefly outlined in an attempt to stimulate thought and further ideas. There are no simple solutions for better thermal technologies. It is hoped that this summary will aid in focusing attention in the critical areas of heat transfer and thermal management for gas turbine hot sections required to make significant new gains in overall efficiency, operability, and durability.

References

[1] Iacovides, Diamnatis Kounadis and Brian E. Launder "Experimental study of the Thermal Development in a Rotating squared-Ended U-Bend. ASME Turbo Expo 2006: Power Land, Sea and Air. May 2006 GT2006-90846

[2] Iacovides, H., Jackson, D.C., Kelemenis, G., Launder, B.E., Yuan, Y.M., "Experiments on Local Heat Transfer in a Rotating Square-Ended U-Bend," International Journal of Heat and Fluid Flow, Vol. 20, pp. 302–310, 1999.

[3] Cheah S C, Iacovides H, Jackson D C, Ji H and Launder B E., "LDA investigation of the flow development through rotating U-ducts." ASME Journal of Turbomachinery, Vol 118, pp. 590–596, July 1996.

[4] Nakamata, C., Okita, Y., Matsuno, S., Mimura, F., Matsushita, M., Yamana, T., and Yoshida, T., 2005, "Spatial Arrangement Dependence of Cooling Performance of an Integrated Impingement and Pin Fin Cooling Configuration," Paper No. GT2005-68348, IGTI Turbo Expo, Reno-Tahoe, Nevada.

Thermal investigation of light emitting diodes

K. Domke & K. Wandachowicz
Poznan University of Technology, Poland

Abstract

High power light emitting diodes are already a popular alternative to incandescent and fluorescent lamps and it seems that in around a dozen years or so they may become the basic type of lamp used in lighting engineering. In addition to emitting light, LEDs also generate a significant quantity of heat. The assumption is that ca. 75% of total power is released in the form of heat which must be carried into the environment. Heat released in the p-n junction of the semiconductor material markedly increases its temperature and thus indirectly affects the photometric and electric characteristics of LEDs. The paper below presents a test stand designed for measuring such characteristics for a wide range of changes of junction temperatures. Junction temperature adjustment can proceed independently of the thermal power released in the LED. Selected measured characteristics of HP LEDs are also presented.

Keywords: light emitting diode, junction temperature, photometric measurement.

1 Introduction

Semiconductor LEDs made in the solid state lightning (SSL) technology can be divided into low power (LEDs) and high power (HP LEDs). The latter represent the most advanced group of light sources used in the lighting industry for general and decorative lighting applications and in the automotive industry for marker lights, stop lights, direction indicator lights and car headlights. The LED market is now a major and rapidly developing segment of the electronics market. Forecast growth of the LED technology (published in 2002) is presented in table 1 (supplemented by 2007 data).

In the LED markets White High Brightness LEDs will fuel growth to surpass $14 billion by 2017. The present LED market by value is as in table 2:

The manufacturing technology of light emitting diodes has been growing exponentially in recent years. Different types and forms of LEDs have varied

WIT Transactions on Engineering Sciences, Vol 61, © 2008 WIT Press
www.witpress.com, ISSN 1743-3533 (on-line)
doi:10.2495/HT080151

electric and photometric parameters. A thorough examination of LED properties is required to construct a luminaire system or vehicle lights with preset input parameters. A factor of basic significance is the relationship between electric/photometric parameters and the working temperature T_j of the p-n junction. Specification sheets for light emitting diodes provide electric parameters (forward voltage U_F) and photometric parameters (luminous flux Φ, colour temperature T_c, chromaticity coordinates xy) for a set value of the forward current I_F in the p-n junction working temperature T_j of 25°C, whereas typical operating temperature T_j of the p-n junction in HP LEDs (with the power value of several watts) exceed 100°C. As the temperature of the p-n junction rises, luminous flux Φ and forward voltage U_F values drop, accompanied by a shift of the wavelength λ_m representing maximum spectral distribution to a higher wavelength. Luminaire manufacturers use a variety of means to reduce the temperature (T_j) of the p-n junction. LEDs are typically installed on heatsinks which, by substantially improving the process of transferring heat generated in the p-n junction into the environment, reduce the temperature T_j of the junction.

Table 1: Forecast growth of the SSL-LED technology according to a report prepared by OIDA (Optoelectronics Industry Development Association) [1–4].

	Estimate		Data	Estimate		Data	Data
	SSL-LED					Incandesc.	Fluoresc.
Parameter	2002	2007	2008	2012	2020	2007	2007
Luminous eff. [lm/W]	25	75	45-100	150	200	16	85
Life time [10^3 h]	20	>20	35-50	>100	>100	1	10
Lumin. flux [lm/lamp]	25	200	-	1000	1500	1200	3400
Power [W/lamp]	1	2.7	-	6.7	7.5	75	40

Table 2: Industrial application LED market 2003–2008 (US$ millions).

Data						Estimate
2003	2004	2005	2006	2007	2008	2017
280	379	425	477	623	817	14 000

2 LED structure

LEDs are p-n junction devices constructed of gallium arsenide (GaAs), gallium arsenide phosphide (GaAsP), gallium phosphide (GaP) and especially HP LED: aluminium indium gallium phosphide (AlInGaP) or indium gallium nitride (InGaN). Silicon and germanium are not suitable because those junctions produce too much heat and no appreciable IR or visible light. The junction in an LED is forward biased and when electrons cross the junction from the n- to the p-type material, the electron-hole recombination process produces some photons in the UV, IR or VIS in a process of electroluminescence. In non-radiative recombination events, the energy released during the electron-hole recombination is converted to phonons. Phonon is a quantised mode of vibration occurring in a rigid crystal lattice that increase temperature and produce heat. A typical design of low and high power light emitting diodes is shown in fig. 1.

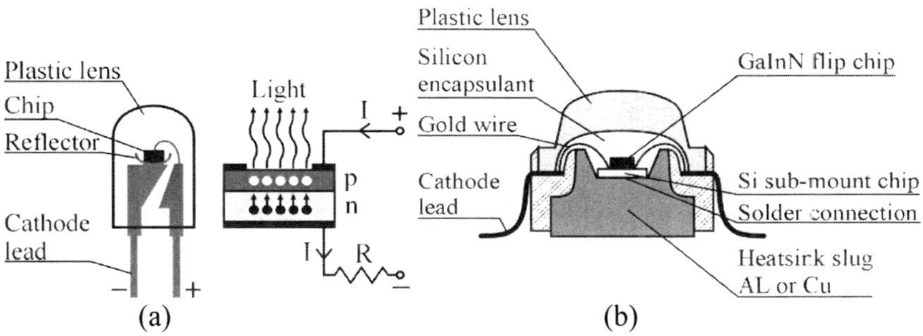

Figure 1: Structure of LED: (a) low power, (b) high power.

Low power diodes are usually made in the through-hole technology and operate without additional heatsink slugs. High power light emitting diodes (HP LEDs) made in the surface mount technology (SMT) have a different design. LEDs of this type, are designed in such a way as to allow easy take away heat from the p-n junction to the metal diode base, as illustrated in fig. 1(b).

HP LEDs are usually placed on a metallic core printed circuit board (MCPCB), often with an additional heatsink slug. Currently manufactured LEDs are generally installed on MCPCB sections and they are assembled in the finished product (e.g. in a luminaire) by mechanical fastening to the heatsink surface, as shown in fig. 2.

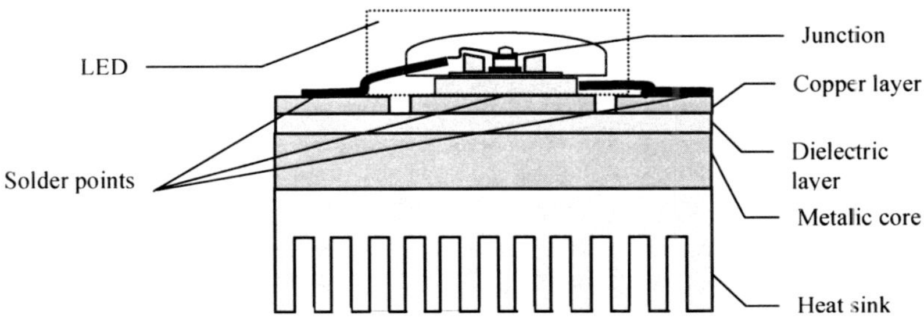

Figure 2: HP LED fastened on MCPCB with a heatsink slug.

3 Heat flow in LEDs

Processes taking place in HP LEDs generate a considerable amount of heat. Most of the heat is released in the p-n junction during the flow of the current I_F. Joule heat generated in voltage conductors supplying voltage to the p-n structure is practically negligible. A characteristic feature to note is the small area (volume of the p-n structure) which produces heat. This results in very high heat power densities and consequent difficulties with dissipating such large amounts of heat. The table 3 is illustrated lists of data for standard currently manufactured LEDs.

Table 3: Standard HP LED specifications [5,6].

LED type	Total power $P=U_F I_F$ [W]	Efficiency η [%]	Heat power $P_{th}=(1-\eta)$ P [W]	Max junc. temp. T_j [°C]	Heat power dens $q=P_{th}/S$ [W/m²]
C460XB900-S92xx-A	1.2	14	0.98	125	$1.4 \cdot 10^6$
EZ1000	3.8	28	2.7	145	$3.1 \cdot 10^6$
OSTAR LEWE3A	27	28	19.4	150	$2.9 \cdot 10^6$

Heat generated in the junction must be carried into the environment. Initially, some heat is dissipated as a result of thermal conduction via internal LED components from the p-n structure area to the external LED components (e.g. to the MCPCB base). A description of the phenomenon is based on the notion of the so-called thermal resistance R_{th}. Thermal resistance Hewitt *et al.* [7] and Wong [8] define as:

$$R_{th,x-y} = \frac{T_x - T_y}{P_{th}} \tag{1}$$

where T_x and T_y stand for *x* or *y* surface temperatures and P_{th} denotes thermal power flowing between those surfaces.

The appropriate formula for calculating thermal resistance R_{th} can be found in literature on thermal conduction in solids [7,8].

In HP LEDs (cf. fig. 1(b)) the direction of the main heat flux is from the p-n junction to the metal heatsink slug and to metal anode and cathode leads soldered to the PCB. The heat flux through non-metallic elements of the diode to the external epoxy housing and the lens accounts for no more than 1% of P_{th} and can be disregarded. Resistances R_{th} which occur in the internal heat flow path are referred to as internal thermal resistances. A connection diagram for thermal resistances inside a LED is shown in fig. 3.

LED internal thermal resistances are difficult to measure and are strictly related to the internal design of specific LED types. Diode manufacturers are striving to work out designs with reduced thermal resistance levels. Reductions in $R_{th,j\text{-}sp}$ levels resulting from modifications in LED design are shown in fig. 4.

Manufacturers generally specify only the total thermal resistance $R_{th,j\text{-}sp}$ measured between the semiconductor junction inside the diode and its metallic core [5,6]. In order to ensure adequate cooling of HP LEDs, the base of the diode is soldered or fastened (e.g. screwed down) to a suitable external heatsink with thermal resistance $R_{th,\ heatsink}$.

Thermal power P_{th} is carried from the heatsink surface: by convection to the gas surrounding the heatsink (with temperature $T_{amb,conv}$) and by radiation to

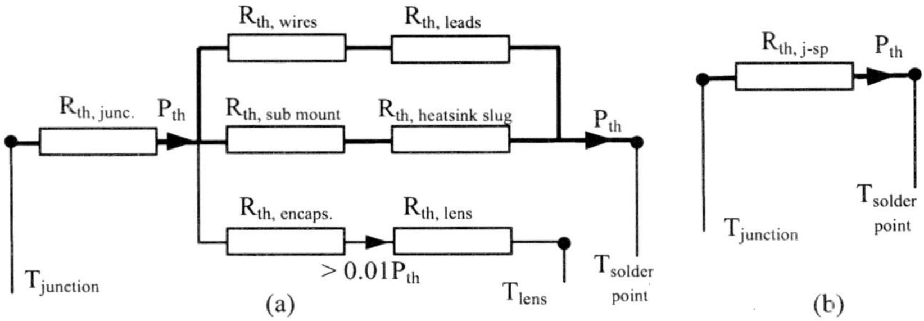

Figure 3: Diagram for internal thermal resistances of LEDs: (a) full, (b) simplified. Main heat flow paths are marked with bold lines. Denotes as in figs. 1–2.

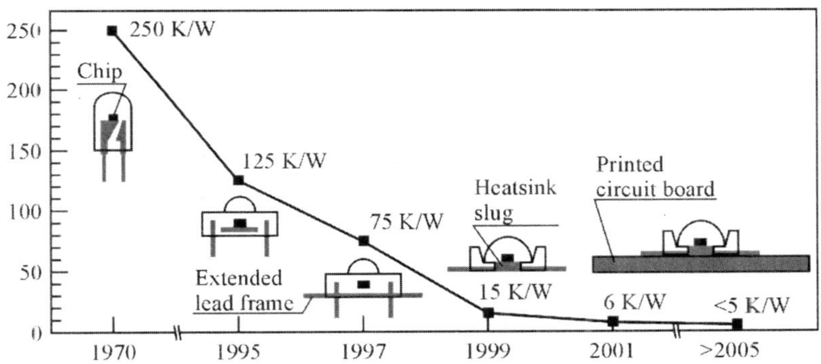

Figure 4: Thermal resistance of LED packages (adopted from Arik *et al.*, 2002).

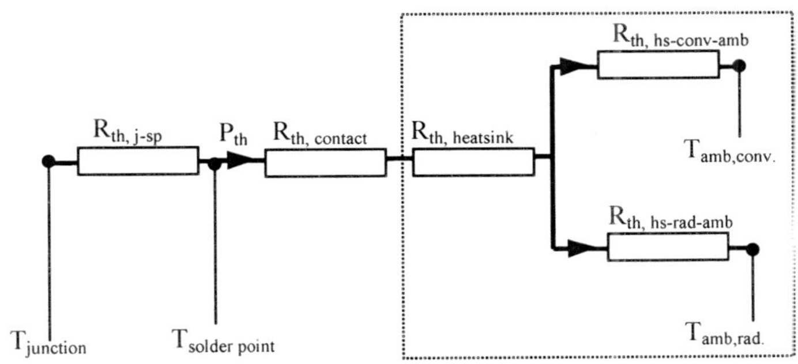

Figure 5: Diagram for thermal resistances of a LED installed on the heatsink.

other bodies surrounding the heatsink (with temperature $T_{amb,rad.}$). Corresponding thermal resistances are marked as ($R_{th,\ hs\text{-}conv\text{-}amb}$) and ($R_{th,\ hs\text{-}rad\text{-}amb}$). In specific cases, temperatures $T_{amb,conv}$ and $T_{amb,rad.}$ can differ. Heatsink manufacturers usually specify total thermal resistance $R_{th,\ heatsink\text{-}amb.}$ and recommend assuming that $T_{amb,}= 0.5(T_{amb,conv}+ T_{amb,rad})$.

The external heatsink is intended to dissipate the thermal power P_{th} from the diode into the environment. It can be calculated using the formula below:

$$P_{th} = (1-\eta)P = \frac{T_j - T_{amb}}{R_{th,tot}} \tag{2}$$

where $P=U_F{\cdot}I_F$ stands for the power of electric losses on the diode, η denotes its efficiency and $R_{th,tot}$ stands for total thermal resistance between the p-n junction and the environment with the temperature T_{amb}:

$$R_{th,tot} = R_{th,j-sp} + R_{th,contact} + R_{th,heatsink-amb} \tag{3}$$

where $R_{th,heatsink\text{-}amb}$ denotes total thermal resistance of the heatsink and $R_{th,contact}$ – thermal resistance of the diode – heatsink connection. The specifications are provided by manufacturers and are the basic factors determining heatsink choice. The formulas given in (2) and (3) show that heatsinks should be selected in such a way as to make the total thermal resistance satisfy the following relationship:

$$R_{th,heatsink-amb} + R_{th,contact} < \frac{T_j - T_{amb}}{P_{th}} - R_{th,j-sp} = \frac{T_{sp} - T_{amb}}{P_{th}} \tag{4}$$

4 Temperature-dependent photometric and electric characteristics. Methods for determining junction temperature

Junction temperature T_j is an important parameter which has a major effect on failure-free LED operation, as well as a range of photometric and electric characteristics. If the permitted temperature level is exceeded, the semiconductor unit is permanently damaged. T_j increasing above T_{amb} causes a deterioration of LED operating characteristics, as illustrated in fig. 6.

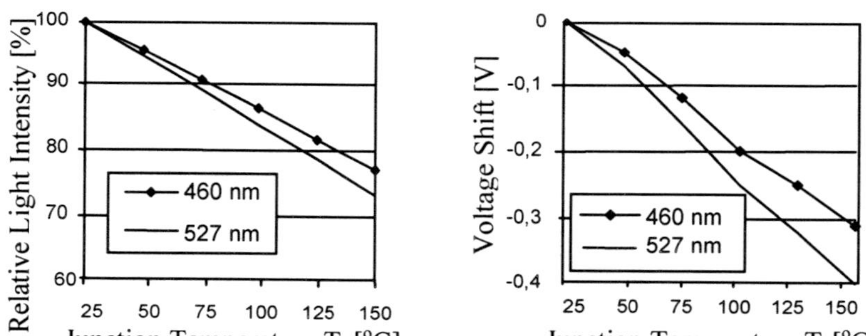

Figure 6: Specifications of EZ1000 LED for I_F = 350 mA. [4]

Unfortunately, junction temperature (T_j) is not a directly measurable quantity. The most frequently used means of determining junction temperature is an indirect method based on determining e.g. the standard relation between the LED forward voltage and junction temperature $U_F=f(T_j)$ in the course of special test measurements. Measurements are carried out after placing the entire LED structure in the heat chamber maintaining constant temperature, adjusting the temperature of the entire semiconductor unit at a uniform level. A momentary current pulse is then applied to the LED. The pulse must be so short as not to produce any thermal effects and, at the same time, make it possible to measure electric quantities. The characteristic $U_F=f(T_j)$ thus obtained is then used to determine junction temperature T_j during measurements of photometric quantities performed as the LED is supplied with direct current in conditions similar to normal LED operations.

The main shortcoming of the method outlined above is the assumption that the $U_F=f(T_j)$ characteristic determined in testing conditions also reflects the relationship in operating conditions, although LED temperature distribution is completely different in both cases.

5 Test stand for performing thermal measurements of LEDs

The basic aim of constructing the test stand was to create appropriate conditions for performing LED tests focused on electric and photometric characteristics in the function of the variable temperature T_j of the p-n junction, with a constant, freely set value of forward current intensity I_F. In order to achieve the objective, it was necessary to come up with a system which, in addition to standard measurements of electric and photometric parameters, would also ensure adjustable inflow or outflow of thermal power P_{th} to the p-n junction.

LEDs selected for testing (fig. 2) are mounted on a metallic MCPCB base with high thermal conductivity λ. The base is fastened to the metal block (Cu), with a Peltier element placed on the opposite surface (see fig. 7).

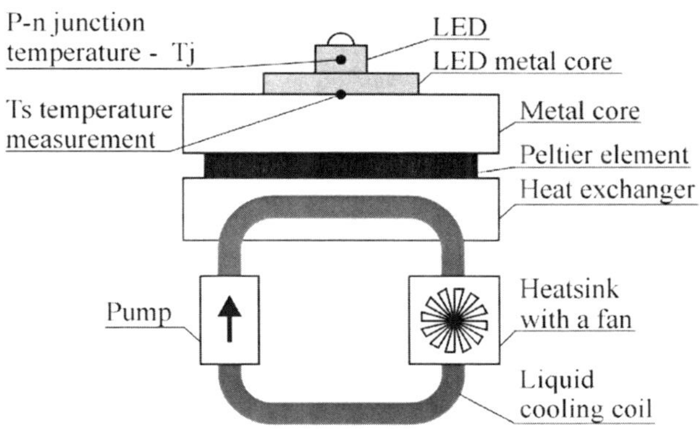

Figure 7: Diagram illustrating the test stand designed for performing tests of thermal characteristics of light emitting diodes (LEDs).

When voltage is applied the Peltier element, the current I_{Plt} is generated, producing a temperature difference which is conducive to the flow of thermal power P_{th} from the LED to the heat exchanger. Therefore, the Peltier element acts as a pump with adjustable delivery rate, transferring thermal power from the LED to the heat exchanger. A connection diagram of thermal resistances of the test stand is presented below:

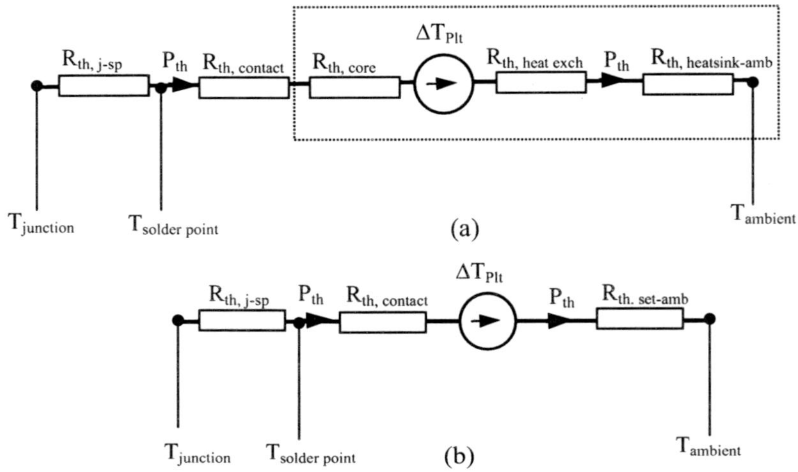

Figure 8: Diagram for thermal resistances of the test stand: (a) full, (b) simplified.

In fig. 8, the Peltier element is shown as a source of temperature difference ΔT_{Plt} (equivalent to the source of voltage) produced by the flow of current with the current intensity I_{Plt}. ΔT_{Plt} can be adjusted by changes in the current intensity I_{Plt}, while the sign of the temperature difference – its direction. Consequently, by changing the polarity of voltage applied to the Peltier element, you may obtain an effect of p-n junction cooling or heating.

Analysing the thermal circuit given in fig. 8, you derive the following:

$$P_{th} = \frac{T_j - T_{Plt1}}{R_{th,j-sp} + R_{th,contact}} = \frac{T_{Plt2} - T_{amb}}{R_{th,set-amb}} \tag{5}$$

and determining the temperature difference produced on the Peltier element as $\Delta T_{Plt}=T_{Plt1}-T_{Plt2}$, the relationship (6) is derived describing the thermal relations found in the analysed test stand.

$$T_j = P_{th}(R_{th,j-sp} + R_{th,contact} + R_{th,set-amb}) - \Delta T_{Plt} + T_{amb} \tag{6}$$

Analysing the formula in (6), it follows that it is possible to obtain a preset junction temperature at any power P_{th} by selecting an appropriate value of $\Delta T_{Plt}=f(I_{Plt})$. The temperature is obtained without interfering with the power P_{th}

generated in the diode. Given the sufficient efficiency of the Peltier element, it is possible to achieve junction temperatures that are lower than the ambient temperature. The test stand described here thus makes it possible to obtain preset junction temperatures for a broad range of changes of thermal power generated in the LED. P_{th} and T_j can be adjusted independently.

The temperature of the junction T_j was determined indirectly, using the formula :

$$T_j = R_{thj-sp} P_{th} + T_{sp} \tag{7}$$

where temperature T_{sp} was measured in the central point at the contact surface of the LED's MCPCB with the metal base of the test stand (cf. fig. 7). Measurements were performed using a K thermocouple (Ø 0.2 mm) in the steady thermal state. The value of thermal resistance R_{thj-sp} was assumed according to the manufacturer's specifications. Thermal power was defined on the basis of measurements of LED electric parameters. The use of the formula given in (7) to determine the junction temperature Tj is recommended by manufacturers of light emitting diodes. As a consequence, it is assumed that thermal resistance R_{thj-sp} was determined for total power supplied to the diode ($P_{th}=U_F*I_F$).

6 Thermal photometric characteristics

Measurements were carried out for two types of light emitting diodes [2, 3] (table 4). The test stand made it possible to obtain a wide range of temperatures to measure temperature characteristics of the examined diodes.

Table 4: Catalogue specifications of light emitting diodes: maximum permitted levels are given in brackets [5,6].

No	LED type	I_F [mA]	U_F [V]	Φ [lm]	T_j [^{0}C]	R_{thj-sp} [K/W]
1	K2 Star L2K2-MWW4	1000 (1500)	3,72	100	110 (150)	13
2	Ostar LE W E3A	700 (1000)	20,8	240 ÷ 520	150 (180)	3

Investigation in the proposed measurement system give the lowest (-9.1°C – K2 Star, 1.2°C – Ostar) and the highest temperatures of the solder point T_s (119°C – K2 Star, 135.1°C – Ostar) and the p-n junction temperature T_j (45.5÷167.8°C – K2 Star, 42.3÷174.1°C – Ostar).

Sample results showing changes of the luminous flux Φ and the forward voltage U_F in the function of junction temperature are given in fig. 9. The characteristics corroborate the relationship between the luminous flux and the temperature of the junction. For recommended maximum operating temperatures of the p-n junction (110°C and 150°C cf. table 5), flux reduction – in relation to catalogue specifications – is considerable, amounting to 26% (K2 Star) and 32% (Ostar).

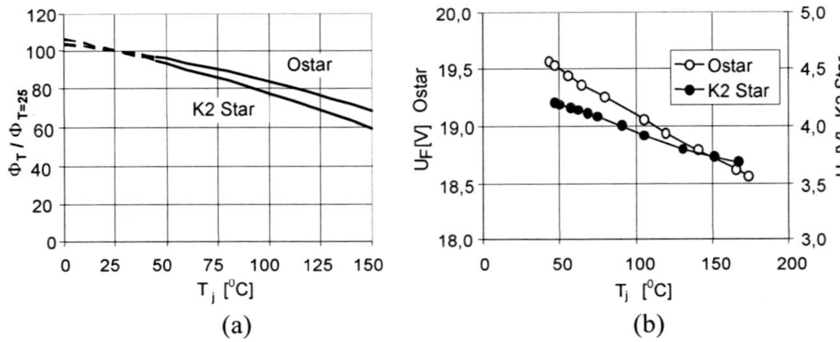

Figure 9: For constant forward current I_F: 1000 mA (K2 Star) and 700 mA (Ostar): (a) Relative changes of the thermal flux and (b) Changes of the forward voltage U_F in the function of the junction temperature T_j.

The extrapolation of measurement curves for low temperature (broken lines in fig. 9) resulted from the impossibility to achieve the assumed temperature of the p-n junction, despite obtaining negative temperature at the solder point T_{sp}. The Peltier element was not adequately efficient and the value of the temperature difference ΔT_{Plt} proved insufficient. The defect will be eliminated in the subsequent version of the test stand.

Figure 9 presents the temperature characteristics of changes in the forward voltage U_F of selected diodes. Within the range of p-n junction temperature changes between 50°C and 150°C, the mean coefficient of forward voltage change was 4.4 mV/°C (K2 Star) and 7.4 mV/°C (Ostar). In view of the typical voltage-current characteristic of LEDs, where minor changes of the forward voltage U_F produce major changes of the forward current intensity I_F, diode supply should be provided in the form of systems of stabilised current supplies, not voltage supplies.

References

[1] Optoelectronic Component Industry to 2008. Reed Electronics Research. http://www.rer.co.uk.
[2] http://www.netl.doe.gov/ssl/PDFs/lifetimeWhiteLEDs_aug16_r1.pdf
[3] Klipstein, D. L. The Brightest and Most Efficient LEDs and where to get them. *Don Klipstein's Web Site*: http://members.misty.com/don/ledx.html
[4] Light Emitting Diodes for General Illumination. An OIDA Technology Roadmap. October 2002.
[5] Cree EZ 000 LEDs. http://www.cree.com/products/pdf/CPR3CR.pdf
[6] Ostar Lightning. Application note. www.osram-os.com
[7] Hewitt G. F., Shires G. L., Bott T.R., Process Heat Transfer, CRC Press, Boca Raton, N. York, 1994.
[8] Wong K-F. V. Intermediate Heat Transfer. Marcel Dekker Inc. N. York, 2003.

Nucleate boiling flow – experimental investigations and wall heat flux modelling for automotive engine applications

H. Steiner[1], B. Breitschädel[2], G. Brenn[1], H. Petutschnig[3] & C. Samhaber[4]
[1]*Institute of Fluid Mechanics and Heat Transfer, Graz University of Technology, Austria*
[2]*Competence Center "The Virtual Vehicle" (vif) Graz, Austria*
[3]*AVL LIST GmbH Graz, Austria*
[4]*BMW Motoren GmbH Steyr, Austria*

Abstract

In the continuing search for highly efficient technical cooling devices, the nucleate flow boiling regime is becoming increasingly appreciated for providing the highest possible heat transfer rates. Regarding the case of liquid cooling systems of combustion engines, it is known that subcooled boiling occurs in thermally highly loaded regions. Therefore, reliable thermal management requires sufficient knowledge on how specific operation conditions and system parameters typically found in engine cooling jackets can affect the subcooled boiling heat transfer. In particular, the present work investigates experimentally the effects of varying volume fractions of the two main components of the coolant, of varying roughness and orientation of the hot wall surface with respect to the vector of gravitational acceleration, and of vibrations of the hot surface. The obtained comprehensive data set is highly valuable to elucidate and quantify the impact of the investigated effects. The data are also used to evaluate the performance of a wall heat flux model, which was specially developed for use in subcooled boiling flow in automotive applications. The model is proven to provide good overall accuracy for the considered experimental conditions.

Keywords: subcooled boiling heat transfer, heated surface roughness, heated surface vibrations, heated surface orientation, liquid composition.

WIT Transactions on Engineering Sciences, Vol 61,
www.witpress.com, ISSN 1743-3533 (on-line)
doi:10.2495/HT080161

1 Introduction

Wherever high thermal power has to be transferred across confined areas, the nucleate boiling flow represents an attractive regime for providing high heat transfer rates. The flow boiling heat transfer is therefore of great interest in various technical applications ranging from electronic chip cooling to high efficiency compact heat exchangers. The particular case of subcooled boiling flow, where nucleate boiling occurs only in a thin superheated near-wall layer, while the outer bulk liquid remains below saturation temperature, is also of great relevance in liquid cooling jackets of combustion engines, where it helps to keep the temperatures of the heated walls on acceptable levels. A more extensive use of the great potential of nucleate boiling requires a sufficient knowledge on how and to what extent specific system parameters and/or operation conditions may affect the nucleate boiling heat transfer. In particular, the present work experimentally investigated the effects of the composition of the working liquid, and the effects of the heated surface roughness, orientation and vibration. The comprehensive experimental data base obtained in these experiments is also highly valuable for the validation of the modelling of the wall heat flux. Accordingly, the present study assesses the accuracy and the limits of a typical ansatz known as Boiling-Departure-Liftoff (BDL) model, which has become a well-established approach for the modelling of the total wall heat flux with subcooled boiling in automotive coolant flow.

2 Experimental setup

The present experimental apparatus basically consists of a closed loop, where the flow is generated by a pump. The working fluid is preconditioned to enter the test section at a defined bulk temperature, velocity and operating pressure. The velocity of the bulk flow can be varied within the range of $0.05 \leq u_b \leq 2.0$ m/s, corresponding to the range of Reynolds numbers between 2200 and 88000. The absolute operating pressure can be set within the range $1.0 \leq p \leq 2.0$ bar. The key part of the device, i.e. the test section, is schematically shown in fig. 1. It is basically a duct with a square-shaped cross section of dimensions 36 x 36 mm^2. The heat flux into the channel is generated by electric cartridge heaters (Bach Resistor Ceramics) located at the bottom of the aluminium heater, from where the heat is conducted to the top of the heater. At the upper surface, whose length is 65 mm and width is 10 mm, the heat is transferred to the working fluid flowing through the channel. The wall temperature as well as the wall heat flux are determined based on measurements of the temperature using twelve K-type thermocouples appropriately distributed in the solid heater. The base plate of the test section, where the top of the aluminium heater is integrated, is made of Polytetrafluoroethylene (PTFE), whose very low thermal conductivity (λ_{PTFE}= 0.23 W/mK) should guarantee lowest possible heat losses from the heater to the surrounding structure. Due to thermal durability restrictions of the PTFE base plate, the maximum heater surface temperature was limited to T_w = 160°C. Glass windows are embedded in the three unheated walls of the test section to make the section optically accessible. For the investigation of the effect of the orientation of the heated surface with respect to the direction of the gravitation force the whole test section could be rotated around the channel axis by a prescribed angle κ. The rotation of the test section was possible while the system remained filled with liquid, so that the heater surface was always wetted to ensure good reproducibility of the results. When investigating the effect of heated surface vibration, the bottom of the aluminium heater was connected to the actuator head of an electro-dynamic shaker (Brüel and Kjaer 45N) to excite a predefined oscillatory motion of the heater. The oscillatory motion was optically recorded with a laser vibrometer.

Using the present experimental setup, errors in the experimentally obtained heat fluxes are mainly due to measurement and position errors of the thermocouples, as well as the uncertainties in the actual thermal conductivity of the heater material (aluminium alloy) and in the heat losses to the surroundings of the heater. A worst-case estimation yielded a total error of the heat flux ranging from ±5% in the convective regime to ±2% in the nucleate boiling regime, always referring to the value actually obtained from the measurements. The error in the measured surface temperatures amounts to ±0.15°C. The inductive flow meter measures the flow rate with a relative error as low as ±0.5% of the displayed value.

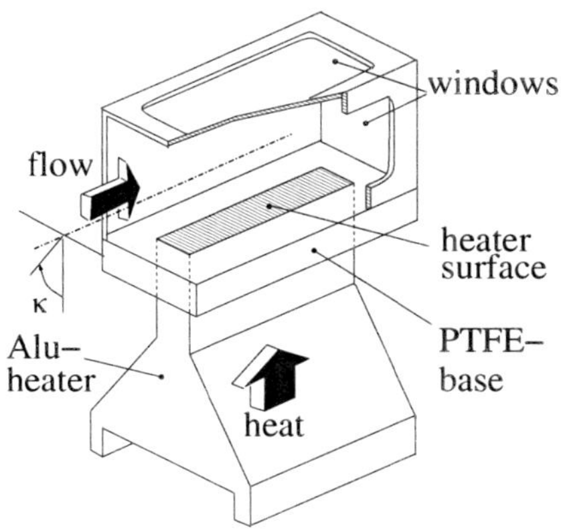

Figure 1: Experimental test section.

3 Examined model for the total wall heat flux

The present work considers the so-called Boiling-Departure-Lift-off (BDL) model, which is basically a linear superposition approach originally introduced by Chen [1]. As such the BDL model assumes the total wall heat flux q_w to be composed of a convective (q_{conv}) and a nucleate boiling contribution (q_{nb})

$$q_w = q_{conv} F + q_{nb} S. \quad (1)$$

The convective component q_{conv} is obtained from the Dittus-Boelter correlation for forced convective channel flows. The factor F represents the enhancement of the convective component due to the bubble agitation. Since in subcooled boiling flow the vapour fractions are generally low, it is argued that this effect is negligible, assuming F=1. The nucleate boiling component q_{nb} is obtained from a classical pool boiling correlation due to Forster and Zuber [2]. The factor S is introduced to reflect the suppression of nucleate boiling observed at increasing flow rates. The BDL approach models S based on a balance of forces acting on the vapour bubbles at the instants of their detachment and lift-off from the heated surface. As such, S is obtained as essentially dependent on the near-wall velocity field. S is unity in the reference case of pool boiling with zero flow velocity, and it decreases to zero, $S \rightarrow 0$, as the bulk flow velocity becomes higher.

The BDL model was originally devised and calibrated for subcooled boiling in automotive engine coolant flow in automotive engines (Kobor [3]). It was later extended to boiling with pure water by Steiner *et al.* [4], where a detailed formulation of the model equations is given. The BDL model has become a well-established approach especially in the automotive industry. The accuracy of the BDL model shall be assessed using the original set of the model coefficients of Kobor [3] when applied to the coolant flow under the particular conditions investigated in the experiments.

4 Results for the investigated effects

The baseline experimental conditions of the present investigations were specified to meet typical operating conditions in automotive liquid cooling systems. Following this premise, the temperature of the bulk liquid at the inlet of the test section was always set to T_b=95°C, and the absolute pressure of the system was was varied within $1.2 \leq p \leq 2.0$ bar. The wall heat flux varied within the range $0 \leq q_w \leq 0.5$ MW/m^2, yielding wall superheats up to $\Delta T_{sat} = T_w - T_{sat} = 40$ K. The considered velocities of the bulk liquid were chosen within $0.05 \leq u_b \leq 0.5$ m/s.

4.1 Composition of the liquid

The considered working liquid basically consists of two main components, deionized water and glysantin G48 from BASF. The latter basically consists of ethylene-glycol and additives, which represents a typical antifreeze component of engine coolants and is widely used in the automotive industry. The present experiments investigated the effect of the binary composition by considering three volumetric mixing ratios, 40/60, 50/50, and 60/40 vol% glysantin/water, respectively.

Two sets of flow boiling curves, which were experimentally measured for the three considered mixtures at the Reynolds numbers Re_b = 5500 and 22000, are shown in figs. 2a und b, respectively. The bulk velocities corresponding to the two Reynolds number are about u_b = 0.1 and 0.4 m/s, respectively. It becomes obvious that in both cases the 60/40Vol% mixture, i.e., the liquid with the smallest fraction of the more volatile component (water), exhibits the lowest total wall heat fluxes at a given wall superheat $\Delta T_{sat} > 0$ in the nucleate boiling regime. This observation is confirmed by previous studies with binary and ternary systems in literature (Kandlikar [5], Wenzel *et al.* [6]), who attributed the observed tendency to a basically non-azeotropic behaviour of the mixture: the fraction of the more volatile component decreases faster in the liquid next to the bubble surface, which increases the effective saturation temperature of the local mixture being enriched with the less volatile component. The boiling curves obtained with the BDL model, also shown in figs. 2a,b, exhibit the experimentally observed tendency as well, even though the predicted heat flux levels are generally somewhat too low. Nonetheless, it is evidently sufficient to model the effect of the varying glysatin/water ratio in terms of a corresponding variation of the material properties of the mixture, which basically means assuming the binary system as an azeotropic mixture, as it is done in the BDL model. The model neglects any non-azeotropic effects, and it computes the material properties based on phase and thermodynamic equilibrium assumptions for a given mixture.

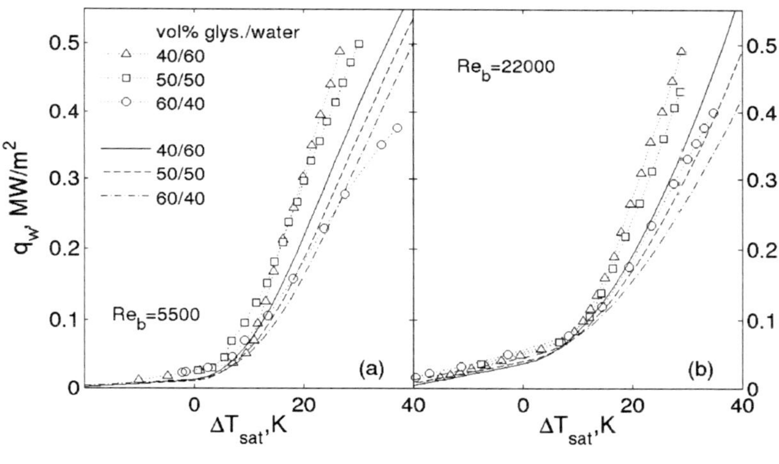

Figure 2: Flow boiling curves for varying coolant compositions: (a) Re_b=5500; (b) Re_b=22000; experimental data denoted by symbols, BDL model predictions by lines; pressure: p=1.5 bar.

4.2 Roughness of the heated surface

It is generally accepted that surface roughness can increase the boiling heat transfer rates only if the roughness elements provide additional active nucleation sites. To remain active they must be capable to trap a gaseous rest after bubble detachment to provide a nucleus for the formation of the next bubble. Thus, the initial effect of surface roughness may finally vanish due to a successive flooding of active cavities and/or a long-term change of the surface microstructure caused by corrosion or deposition of solids from the liquid (these processes are frequently subsumed as "aging"). Such a levelling down of primary roughness effects was already observed in an early work by Jakob and Fritz [7] for boiling of water on copper plates with different surface finish. The present experimental investigations consider three types of surface roughness, whose mean (R_z) and maximum (R_{max}) peak-to-valley roughness heights are listed in tab. 1.

Table 1: Roughness heights of heated surface types.

Surface	R_z,µm	R_{max},µm
Aluminium as cast: 'standard'	45.7	51.7
Aluminium as cast: 'rough'	130	143
Aluminium polished: 'smooth'	1.99	2.38

The first surface termed 'standard' matches the surface characteristics of the cast aluminium material commonly used for automotive cylinder heads. As such, it represents a reference case. The second surface termed 'rough' was produced by sand casting, using very coarse grains leading to big superficial cavities. The third surface termed 'smooth' is originally a sand casted standard surface, which has been milled and polished. The experimentally measured flow boiling curves, as obtained after increasing operation time, are shown in fig. 3.

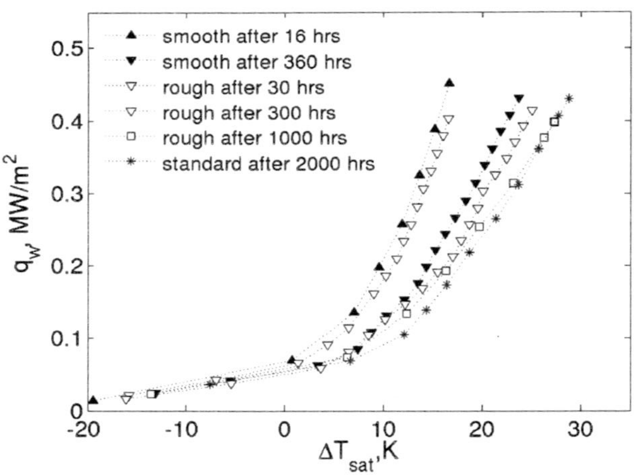

Figure 3: Flow boiling curves for varying surface roughness after different operation times; velocity of the bulk liquid u_b=0.476m/s; mixture 50/50Vol% glysantin/water, pressure p=1.5 bar.

It is seen that both the smooth and the rough surfaces exhibit almost the same heat transfer conditions at a primary stage, since the earliest measured boiling curves of both surfaces almost coincide. The considerable concentration of large cavities on the very rough surface evidently does not provide additional active nucleation centres compared to the smooth polished surface. With increasing operation time, both types of surfaces exhibit a notable degradation of the heat transfer rate, as seen from the shift of the boiling branches towards higher wall superheats. They finally approach the boiling curve of a long-term measurement using the standard surface. The present results are evidently well in line with the aforementioned observations of other authors in literature: the aging related changes of the microgeometry during long-term operation level out any potential primary surface roughness effect on the boiling heat transfer. For the presently considered surface types the effect of surface roughness can be therefore neglected. Accordingly, it need not to be explicitely considered in the modelling of the wall heat flux either.

4.3 Orientation of the heated surface

Due to the large liquid/vapour density ratio, buoyancy, and hence the orientation of the heated surface with respect to the gravitational vector, may strongly influence the bubble dynamics in boiling flow. The present work investigated in particular three cases of the surface orientation, an upward facing surface heated from below, representing the reference case, a horizontally sideward facing surface heated from aside, and a downward facing surface heated from above. The flow boiling curves measured for different velocities of the bulk liquid u_b did not differ notably for the upward and the sideward facing cases. Contrarily, the downward facing case exhibited a rapid increase of the wall superheats at certain wall heat fluxes in the low velocity range, as it is seen in figs. 4a-c. The optical observation of the conditions on the heated surface clearly attributes this critical decrease of the wall heat transfer rate to significant bubble agglomeration leading to the formation partial vapour films. Based on a dimensional analysis, a criterion for the observed transition from nucleate to partial film boiling was derived in terms of the non-dimensional correlation

$$Bo_{trans} = C\,Re_L\;Ec_{sub}^{-n}\,. \qquad (2)$$

This correlation involves the Boiling number $Bo_{trans}=q_{w,trans}/u_b\rho_l h_{lg}$, the Reynolds number $Re=u_b\rho_l L/\eta_l$ with the length scale $L=[\sigma_{lg} / g\,(\rho_l-\rho_g)]^{0.5}$, and the Eckert number $Ec_{sub}= u_b^2/c_{p,l}(T_{sat}-T_b)$, where $\rho_l,\eta_l,c_{p,l}$ are the density, dynamic viscosity, specific heat of the bulk liquid, respectively, σ_{lg} the surface tension, and h_{lg} the latent heat. $C=8.5\cdot10^{-11}$ and n=- 0.63 are empirically determined model parameters. As shown in figs. 4a-c by the dumbbell-shape marks, the transitional heat fluxes $q_{w,trans}$ obtained from the correlation given in eq. (2) basically demarcate the transition from nucleate to partial film boiling. As such, they also represent a reliable upper limit for the applicability of the BDL model, whose predictions start to deviate significantly from measurements for wall heat fluxes q_w exceeding $q_{w,trans}$

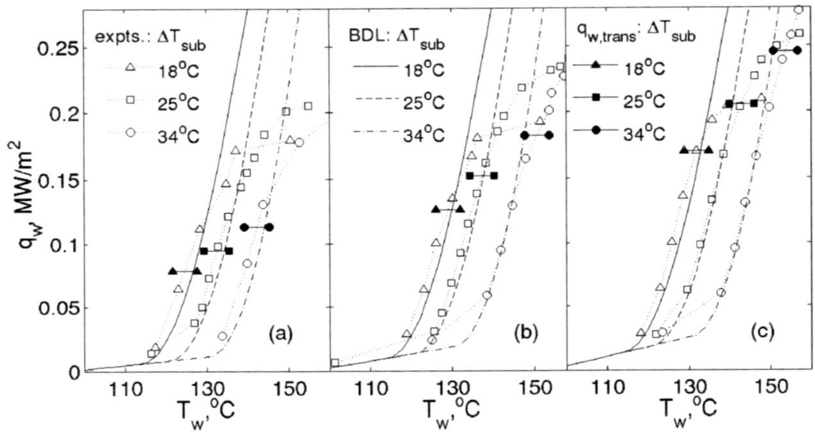

Figure 4: Flow boiling curves with downward facing surface for different subcoolings $\Delta T_{sub}=T_{sat}-T_b$ and varying velocities of the bulk liquid: (a) u_b =0.05m/s, (b) u_b =0.095m/s, (c) u_b =0.142m/s; mixture 50/50Vol% glysantin/water, pressures for the three increasing subcoolings: p=1.2, 1.5 and 2.0 bar, respectively.

4.4 Vibrations of the heated surface

Previous studies in literature on the effect of heated surface vibrations on the heat transfer rate have mainly considered the pool boiling case (see, e.g., Bergles [8]). The present study considers flow boiling with vibrations of the heated surface typically occurring in automotive liquid cooling systems. The approximately sinusoidal surface motion, which is excited by the shaker in the experiments, is parameterized in terms of the frequency f and the rms-velocity of the vertically oscillating surface

$$v_{vib,rms}=\left[\frac{1}{\tau}\int_{t=t_0}^{t_0+\tau} v_{vib}^2(t)dt\right]^{1/2}, \quad \tau\rightarrow\infty, \tag{3}$$

with $v_{vib}(t)$ obtained from the laser vibrometer. The investigated oscillation conditions are within the range of $0.0\leq f\leq 1000$Hz and $0\leq v_{vib,rms}\leq 0.02$m/s. The measurements for the considered oscillatory conditions revealed that the influence of the oscillations on the heat transfer is very small and that it is limited to very low velocities of the bulk. This is exemplarily shown in fig. 5a, where a somewhat higher heat transfer rate is observed at the higher wall heat fluxes in the oscillating case. Fig. 5b shows the results of more detailed measurements carried out at a fixed high and a low heat flux level, denoted $q_{w,H}$ and $q_{w,L}$, respectively, for varying frequencies and rms-vibration velocities. At the high heat flux, the wall temperatures tend to get a little bit lower as $v_{vib,rms}$ increases, which indicates an increasing heat transfer rate. The low heat flux case shows the opposite tendency. Since the here observed quantitative variation of the wall superheat is still very small in both cases, it is concluded that, under the considered oscillatory conditions relevant in automotive cooling, the effect of surface vibrations on heat transfer is negligible. Hence, it need not be accounted for in the wall heat flux modelling either.

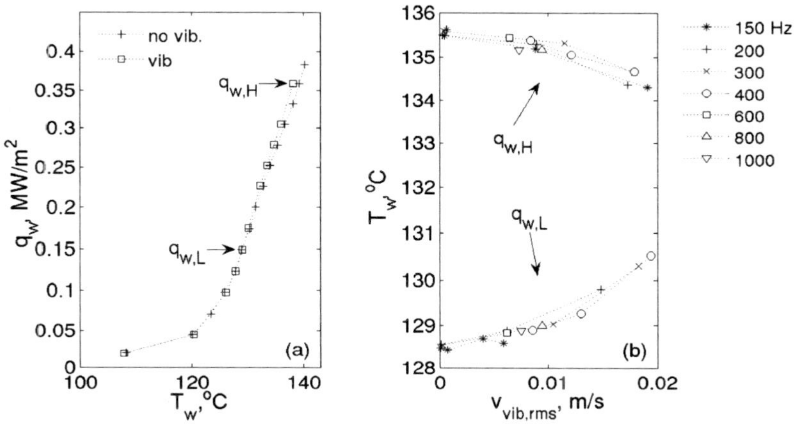

Figure 5: (a) Flow boiling curves with heated surface vibrations (f=200Hz, $v_{vib,rms}$=0.015m/s) and without vibrations (f=0, $v_{vib,rms}$=0); velocity of the bulk liquid u_b =0.05m/s; (b) wall temperatures at two wall heat fluxes $q_{w,H}$ and $q_{w,L}$ for varying oscillatory conditions (f, $v_{vib,rms}$) of the surface; mixture 50/50Vol% glysantin/water, pressure p=1.2 bar.

5 Conclusions

The present study experimentally investigated various effects on the subcooled flow boiling which are of relevance in automotive liquid cooling systems, where the highly efficient flow boiling heat transfer is becoming increasingly important for a reliable and save thermal management. The measured data are also used to evaluate a popular approach for modelling the wall heat flux, known as the BDL model.

The investigations on the influence of the mixing ratio of the two main components of the coolant show that the heat transfer rates in the boiling regime tend to get lower when the fraction of the more volatile component is smaller. The tested wall heat flux model, which basically assumes the coolant as an azeotropic mixture, reflected the observed tendency very well. This suggests that the effect of the mixing ratio can be captured with sufficient accuracy in terms of the material properties of the mixture, neglecting non-azeotropic effects.

The relevance of the effect of the surface roughness turned out to be very limited in time. Long-term operation always produced a microstructure on the surface, which finally lead to approximately the same boiling behaviour of all investigated surface finishes. Based on this observation it is suggested to disregard the effect of the surface finish in terms of a roughness height in the wall heat flux model.

The investigations of effects of the surface orientation demonstrated that, in the case of a downward facing surface heated from above, a transition from nucleate boiling to partial film boiling can occur at low velocities of the bulk liquid. A non-dimensional criterion is proposed to provide a reliable estimate for the wall heat fluxes, where the critical transition from nucleate boiling to partial film boiling has to be expected. Below the so estimated transitional wall heat fluxes, the BDL wall heat flux model showed good agreement with the experiments without any modifications to account for the effect of orientation.

For the considered oscillatory conditions relevant in automotive cooling systems, a very small effect of heated-surface vibrations on the flow boiling heat transfer was observed, occurring only at very low velocities close to the pool boiling limit. It is therefore practically negligible in the wall heat flux model as well.

Acknowledgements

Financial support of this work from the K plus Competence Center Program initiated by the Austrian Federal Ministry of Transport, Innovation, and Technology (BMVIT) and funded by FFG, Land Steiermark, and Steirische Wirtschaftsförderung (SFG) is gratefully acknowledged. We are also grateful for financial support from AVL List GmbH Graz and BMW Motoren GmbH Steyr.

References

[1] Chen, J.C., A correlation for boiling heat transfer to saturated fluids in convective flow, *ASME preprint 63 HT34 presented at the 6th National Heat Transfer Conference*, Boston, 1963.

[2] Forster, H.K. & Zuber N., Dynamics of vapor bubbles and boiling heat transfer, *AIChE Journal*, **1**, pp. 531–535, 1955.

[3] Kobor, A., *Entwicklung eines Siedemodells für die Simulation des kühlmittelseitigen Wärmeübergangs bei Verbrennungskraftmaschinen*, Doctoral thesis, Graz University of Technology, 2003.

[4] Steiner, H., Kobor, A. & Gebhard, L., A wall heat flux model for subcooled boiling flow, *International Journal of Heat and Mass Transfer*, **48**, pp. 4161–4173, 2005.

[5] Kandlikar, S.G., Boiling heat transfer in binary systems: Part II – flow boiling, *ASME Journal of Heat Transfer*, **120**, pp. 388-394, 1998.

[6] Wenzel, U. & Müller-Steinhagen, H., Heat transfer to mixtures of acetone, isopropanol and water under subcooled flow boiling conditions – I. Experimental Results, *International Journal of Heat and Mass Transfer,* **37**, pp. 175–184, 1994.

[7] Jakob, M. & Fritz, W., Versuche über den Verdampfungsvorgang, *Forschung auf dem des Gebiete Ingenieurwesens,* **2**, pp. 435-447, 1931.

[8] Bergles, A.E., The influence of heated-surface vibrations on pool boiling, *ASME Journal of Heat Transfer*, **91,** pp. 152-153, 1969.

Air curtains of open refrigerated display cases revisited: a new technique for infiltration rate measurements

M. Amin[1], H. K. Navaz[2], D. Dabiri[1] & R. Faramarzi[3]
[1]Aeronautics and Astronautics Engineering Department, University of Washington, WA, USA
[2]Mechanical Engineering Department, Kettering University, MI, USA
[3]Refrigeration and Thermal Test Center, Southern California Edison Company, CA, USA

Abstract

Air curtains are created in open refrigerated vertical display cases for creating an invisible barrier between the cold air inside and the warm air outside the case. A systematic approach is developed to minimize the entrainment and infiltration of warm air into the case by optimizing the performance of an air curtain through adjusting pertinent flow parameters and case geometry. A modular display case was manufactured for the parametric studies. In this modular display case the geometry and flow parameters were changed and the infiltration rate was measured using a new technique, tracer gas method, which could be performed noticeably faster than the conventional methods. This apparatus is referred to as the proof-of-concept air curtain (POCAC). A matrix of all possible permutations was constructed with the infiltration rate being the outcome of this matrix. This matrix was populated by actual experimental measurements as well as using validated Computational Fluid Dynamics (CFD) computer programs as a tool to acquire better resolution of the input and output datasets. All the problem variables are referred to as the input vector of all parameters that can be altered, and the output is the infiltration rate. An artificial neural network (ANN) program was used to provide the linkage between the input vector (problem variables) and the problem outcome (infiltration). This program can also be used by industry as a tool to estimate the infiltration rate for all existing open vertical display cases.

Keywords: air curtain, open vertical refrigerated display case, infiltration, entrainment, tracer gas, neural network, experiment, CFD, neural network.

Nomenclature

A a constant
B a constant
CFD Computational Fluid Dynamics
C mass concentration of tracer gas $\left(\frac{\dot{m}_{trac}}{\dot{m}_{trac}+\dot{m}_{air}}\right)$
D average distance between the shelves
DAG Discharge Air Grille
Gr Grashof number ($\frac{G\beta_{\circ}(T_{DAG}-T_r)H^3}{\upsilon}$)
H opening height
I average turbulence intensity at the DAG
L length of the display case in z-direction
$\dot{m}$ mass flow rate
$N.I.R.$ Non-dimensional Infiltration Rate
RAG Return Air Grille
Re Reynolds number (Vw/υ)
$R.H.$ Relative Humidity
Ri Richardson number (Gr/Re)
T average temperature
V average normal velocity
w width
X average horizontal distance between perforated back panel and DAG
Y horizontal offset distance between DAG and RAG

Greek symbols

α offset angle $\alpha = \tan^{-1}\left(\frac{Y}{H}\right)$
η thermal entrainment
β throw angle
$\beta_{\circ}$ thermal expansion coefficient
υ kinematic viscosity

Subscripts:

BP Back Panel
DAG Discharge Air Grille
inf infiltrated
Rm room
RAG Return Air Grille
s shelf
tot total flow rate
$trac$ tracer gas

1 Introduction

An air curtain is a planar jet of air with large aspect ratio that has higher momentum than its surrounding air and is mainly responsible for separating two

WIT Transactions on Engineering Sciences, Vol 61, © 2008 WIT Press
www.witpress.com, ISSN 1743-3533 (on-line)

adjacent zones that have different characteristics and properties, e.g. temperature, airborne particle, relative humidity, etc. There are various applications for creating an air curtain such as infant incubators and clean rooms in hospitals, thermal protection of storage rooms and many aspects of the HVAC industry, microelectronics production line, managing fire in tunnels, protecting art work against airborne pollutions, preventing the penetration of warm air into open refrigerated display cases (current work), etc. Air curtains are usually introduced in vertical directions and depending on the application, they may be circulated through the system (current work), or impinged on a surface located in the downstream of the jet. They may also be isothermal or thermally conditioned depending on the specific application. This work focuses on the protection of the food and products on the shelves in an open refrigerated display case schematically shown in Figure 1, and prevents the penetration of the warmer room air into the cold conditioned display case air.

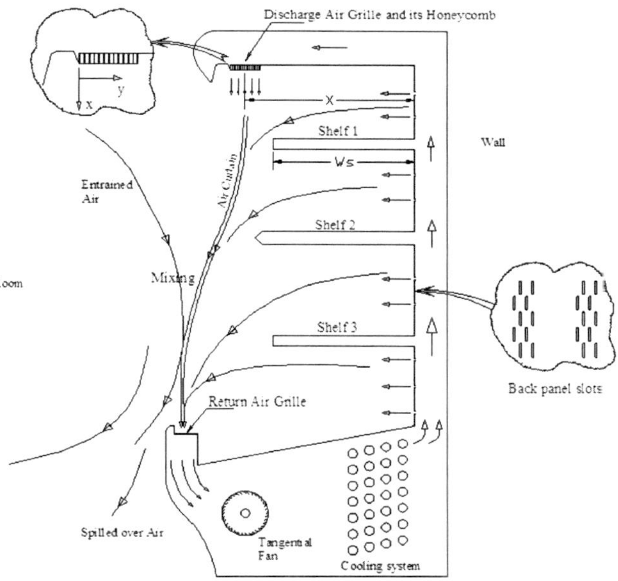

Figure 1: Schematic of a typical display case without food.

The penetration is initiated by entrainment mechanism of the room air due to shear effects into the air curtain. The entrained air with higher content of energy than the supplied air by display case, proceeds in two directions: a portion of that reaches the shelves' products and directly increases their temperature and the rest moves towards the Return Air Grille (RAG), through which the air is circulated. The circulated air that is a mixture of cold and warm air is reconditioned by passing it through the cooling coils of the display case causing the major portion of the cooling load.

One of the early studies on air curtains was performed by Howell et al. [1]. Their work showed that heat and mass transfer through air curtains directly depends on the velocity of air at the discharge air grille (*DAG*). It should be noted that the air curtain is mainly momentum driven causing some shear with the neighboring warm air therefore resulting in mixing. To reduce the amount of mixing, it is important for the air stream to maintain its integrity and path implying that sufficient momentum should exist at the origination point. However, increasing the velocity at the *DAG* will enhance turbulence in the air curtain path as it moves down towards the *RAG*. This increase in turbulence intensity will induce more mixing with the warm air along the main path of the flow, thereby increasing the infiltration rate. In his other work Howell and Shiabata [2] showed that the ratio of the opening height to the width of the *DAG* influences the performance of the air curtains. For a given velocity, the excessive increase of the opening height will cause the air curtain to bend outward and lose its structure and break into a wide region of mixing layers before approaching the *RAG*. On the other hand reducing the opening height will have some practical implications by constraining food accessibility. Howell and Shiabata [2] and Nuygen et al [3] have performed some preliminary numerical studies with turbulence intensity at a free jet *DAG* and its effect on non-circulated air curtains. However, a jet with a circulating air return that can be situated in different locations (in front of the jet or with an offset distance from the jet exit plane) is a more complex problem. The additional complexity can be induced by the geometrical factors, i.e. a free jet is not a bounded domain, while the geometry in a display case resembles a cavity (Rouaud and Havet, [4]). Furthermore, for a non-isothermal flow the air curtain is subject to a transverse pressure difference caused by stack effect (Hayes and Stoecker [5]). Further studies by Howell and Adams [6] revealed that up to *75%* of the refrigeration load of display cases with circulated air curtains could be attributed to the entrainment and infiltration from the room air. In terms of non-dimensional quantities, the above parameters can be grouped as Vw_{DAG}/υ (Reynolds number), I_{DAG}/V, and H/w_{DAG}. Some of the recent studies have focused mostly on the optimization of air curtains in open refrigerated display cases. Several factors such as velocity magnitude, velocity profile shape, and turbulence intensity all at the *DAG,* and some geometrical dimensions of the system including opening height, the horizontal position of the *RAG* relative to the *DAG*, the ratio of the discharged flow through the air curtain to that of the perforated back panel, etc. can affect the performance of the air curtains. Some of these factors have been addressed in the work of many researches, but almost none has investigated the problem in a systematic approach. During the late 90's CFD applications found their way into analyzing mainly the air flow, heating and/or cooling in food industry [7]. A more modern analytical approach that takes advantage of sophisticated tools such as CFD with the same basic goal has been adopted by Axell and Fahlen [8,9]. They developed a correlation for the Nusselt Number of air curtain that related to the amount of heat transfer and cooling load thereafter. Navaz et al. [10] have demonstrated that a marriage between the Digital Particle Image Velocimetry (*DPIV*) and CFD simulation can be quite effective in developing a milestone for systematic study of air curtains. They have implemented the boundary conditions taken from the experimental results and demonstrated that curve fits similar to previous works can be re-produced. Then they used their hybrid CFD/DPIV approach to identify the most important parameters that can impact the performance of an air curtain [11]. They have shown that this hybrid approach can effectively quantify the effect of individual flow/geometrical parameters on the air curtain performance. They have also mentioned that certain operating conditions can result in improved performance of an air curtain [12,13]. Field et al. [14,15] studied circulating air curtains of a display case by the PIV visualization. For better understanding of the buoyancy effect, a flat vertical wall was installed in front of the shelves. They found out that for an isothermal air curtain, the thickness of the jet grows almost linearly downstream and it is somewhat independent of the typical Reynolds numbers. In non-isothermal condition, the negative buoyancy accelerated the jet at low *Ri* values; and this higher momentum causes the spreading of the jet to be delayed. Due to the acceleration, the thickness of the refrigerated curtains was smaller than its isothermal counterpart. Although their work is valuable, they have not considered the effect of the back panel flow, the presence of shelves, and other geometrical factors. The flat wall geometry in their analysis is based on their assumption of the display case to be 100% filled with food products. Chen and Yuaa [16] and Mhiri et al. [17] numerically studied the effect of the Richardson number on the temperature distribution inside cavities exposed to non-circulated air curtains. Chen et al have discussed that for a given Grashof number there is a Richardson number that must be less than a critical value to assure thermal insulation.

Insofar, there are studies focused on different aspects of the air curtain problem. However, to this date there is no systematic approach that can solve industry problems in answering a fundamental question: Under what flow/geometrical parameters an air curtain performance can be optimized, i.e. minimizing the infiltration rate. It is the main purpose of this paper to demonstrate that a hybrid approach of numerical/experimental/ANN can be effectively used to develop a tool for estimating the infiltration rate of the existing display cases, and also find those optimum conditions that can minimize the infiltration rate with constraints imposed by the manufacturer.

2 New approach

In order to achieve the goal of this study that is using a systematic approach to find the minimum infiltration rate in an open refrigerated display case, the most important variables that are shared among all display cases are identified. Basically, all the variables can be divided into two groups, geometric and flow related parameters. By using dimensional analysis the infiltrated mass flow rate can be found to be:

$$\frac{\dot{m}_{\mathrm{inf}}}{\dot{m}_{tot}} = f\left(\frac{L}{w_{DAG}}, \frac{H}{w_{DAG}}, \frac{w_{RAG}}{w_{DAG}}, \frac{w_s}{w_{DAG}}, \frac{X}{w_{DAG}}, \alpha, \beta, \mathrm{Re}, RH, \frac{D}{w_{DAG}}, \frac{T_{DAG}}{T_{Rm}}, \frac{\dot{m}_{BP}}{\dot{m}_{tot}}\right). \tag{1}$$

where $\dot{m}_{tot} = \dot{m}_{RAG} = \dot{m}_{DAG} + \dot{m}_{BP}$.

These parameters are depicted in Figures 1 and 2. In eqn (2) considering that the actual data will eventually be required to be taken for all prescribed number of levels specified for each variable in eqn (1), the number of experiments become prohibitively large and impractical. Therefore, only the most important non-dimensional groups from the above equation are considered. By summarizing eqn (1) we will have:

$$\frac{\dot{m}_{\text{inf}}}{\dot{m}_{tot}} = f\left(\frac{H}{w_{DAG}}, \alpha, \beta, \text{Re}, \frac{\dot{m}_{BP}}{\dot{m}_{tot}}\right) \times \left(\frac{T_{DAG}}{T_{Rm}}\right)^{A} \times (R.H.)^{B}. \tag{2}$$

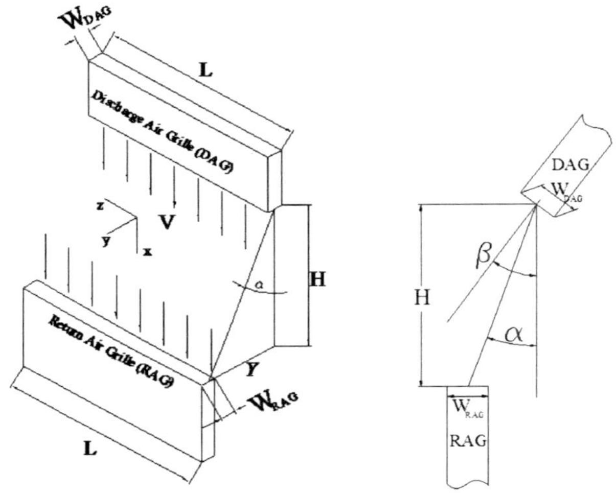

Figure 2: Position of the *RAG* relative to the *DAG*.

The effects of temperatures and relative humidity are presented by a power law with *A* and *B* representing the curve fit constants to be found experimentally. This equation will serve as a guideline to design the modular unit that will be used in our experimental efforts. The overall method can be summarized as: 1) To build a modular air curtain referred to as the proof-of-concept air curtain (POCAC) in which most parameters in eqn (2) can be altered, 2) To measure the infiltration rate directly, 3) To divide the applicable range of each parameter into several intervals and measure the infiltration rate for all permutations, and 4) To use the input and output (infiltration) data for an artificial neural network (ANN) training. The ANN will serve as the tool to not only estimate the infiltration rate of the existing display cases, but also perform parametric studies that will lead to an optimized design. In the current work variables $\frac{H}{w_{DAG}}, \alpha, \beta, \text{Re}, \frac{\dot{m}_{BP}}{\dot{m}_{tot}}$ take 3, 3, 4, 4, and 4 values between their recommended minimums and maximums, respectively. Therefore, a total of 576 combined numerical and experimental results are required to have a reasonable resolution of data.

Table 1 shows the parameters and their corresponding values in their assigned levels. A matrix of test cases was constructed and the data was collected for about one fourth of the total required number. The results were used to train the

ANN program. The ANN prediction for known data provided the level of accuracy. Then the resolution of data was increased by increments of 10-20 experiments until the required accuracy was achieved.

Table 1: Independent variables and their values.

α (degree)	0	16	24	
H/w	8	12	16	
$\dot{m}_{BP} / \dot{m}_{tot}$	(average values)= 0, 0.35, 0.55, 1			
β (degree)	-5	0	5	13
Re_{RAG}	4 values for each "Back Panel flow" ratios, ranging from about 2200 to 12300			

3 Experimental set up

The modular display case (Figure 3) was constructed such that all the variables in eqn (2) can be easily varied. The installed horizontal and vertical bellows, help changing the Y and H, or α and $\frac{H}{w_{DAG}}$, respectively. Furthermore, by using the vertical bellows the discharge duct can rotate to provide the desired throw angle (β). In addition, an attempt was made to lower the turbulence intensity at the discharge of the air curtain to minimize the effect of the turbulence intensity at the jet origination point on the infiltration rate. Thus, to lower the turbulence intensity several flow straightners were inserted inside the air passages. The air flow was supplied by a cross-flow fan that spanned over the entire length (L) of the display case and the velocity of the flow at the discharged was controlled by a DC frequency controller integrated to the fan. The length (L) of the POCAC is *114 cm* and the width of the *DAG* and *RAG* are respectively *4.1 cm*.

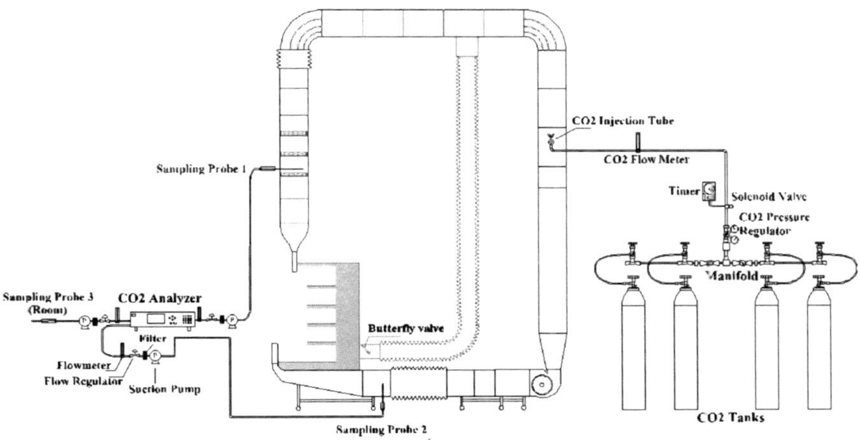

Figure 3: Experimental set up.

4 Infiltration measurement

The conventional experimental method for measuring the infiltration rate in an open refrigerated display case is based on weighting the condensate collected from the cooling coils. Later on, a methodology based on the enthalpy of the room, DAG, and overspill air was developed [11].

However, both methods in an isothermal problem will fail. We have used a tracer gas method to directly monitor the footprint of the infiltrated air through the concentration measurements. In this application, a tracer gas is continuously injected in the upstream and sufficiently far from the *DAG* after the fan. A portion of the tracer gas will reach the *RAG* after it is mixed with outside air and the remainder is spilled into a large room. The concentration of the tracer gas is monitored inside the *RAG* duct (before the fan) and the room. Figure 3 illustrates the measurement equipment and sampling locations. To provide air (including the tracer gas) to the perforated back panel, 3 ducts were used after the injection point of the tracer gas to branch off a portion of the total flow towards the back panel. The flow rate through the perforated back panel was varied by valves. Amin et al [18] have analytically demonstrated that a relationship between the tracer gas concentrations and the infiltration rate can be established as:

$$\dot{m}_{\text{inf}} = \dot{m}_{RAG} \times \left| \frac{C_{DAG} - C_{RAG}}{C_{DAG} - C_{Rm}} \right|. \tag{3}$$

This equation indicates that the absolute amount of infiltrated mass depends on the concentration of the tracer gas wherever it exists and the total flow rate of the display case, i.e. the *RAG* flow rate. Therefore, it is required to measure the concentration of the tracer gas at three locations: 1) upstream of the *DAG* after the injection point, 2) downstream of the *RAG* before the fan and injection point, and 3) the room. To express the infiltration rate as a non-dimensional scalable quantity and also to make the output compatible with the ANN algorithm (all variables must be between 0 and 1) a non-dimensional Infiltration Rate (*N.I.R.*) can be written as in eqn (4).

$$N.I.R. = \frac{\dot{m}_{\text{inf}}}{\dot{m}_{RAG}} = \left| \frac{C_{DAG} - C_{RAG}}{C_{DAG} - C_{Rm}} \right|. \tag{4}$$

This equation resembles the thermal entrainment used in some previous studies [19] given by:

$$\eta = \left| \frac{T_{DAG} - T_{RAG}}{T_{DAG} - T_{Rm}} \right|. \tag{5}$$

In the limiting case, when the entire discharged tracer gas returns to the return duct ($C_{DAG}=C_{RAG}$ *in* eqn (4)) the infiltration goes to zero, that is consistent with $T_{DAG}=T_{RAG}$ and $\eta \rightarrow 0$ in eqn (5). Similarly if no portion of the air from the *DAG* reaches the *RAG* (for instance when the air curtain breaks due to bending outward), the *RAG* concentration and temperature become identical to that of the room assuming that the room is a large reservoir. Therefore, if $C_{Rm}=C_{RAG}$ or similarly $T_{Rm}=T_{RAG}$, they will yield equivalent results, i.e. $N.I.R. \rightarrow 1$ and $\eta \rightarrow 1$.

5 Results

The velocity profile at the *DAG* for the POCAC was measured and visualized by the DPIV method. The desired Top-Hat profile was observed and a turbulence intensity of about 2% was measured. Figure 4 shows the velocity profile at different Reynolds numbers. It is shown [12] that the Top-Hat profile is the best practical profile that can exist at the *DAG* for reducing the infiltration rate. Carbon dioxide (CO_2) has been used as the tracer gas in the current work. Its density is comparable to that of air and it is colorless, odorless, non-corrosive, not flammable, and not hazardous for up to 5000 ppm (for 8 hours of continuous exposure in the room) which is about 15 times more than its amount in nature (350 ppm). The maximum amount of the injected CO_2 is about 25000 ppm (equivalent to 2.5% mass fraction) before the *DAG* that falls under 5000 ppm (0.5%) in the room and across the air curtain. Figure 5a shows the concentration of the tracer gas at the three points of interest in a typical experiment including the N.I.R. It can be seen that the N.I.R. remains constant even after the injection of the tracer gas has stopped. The results are for steady state operation of the display case and air curtain.

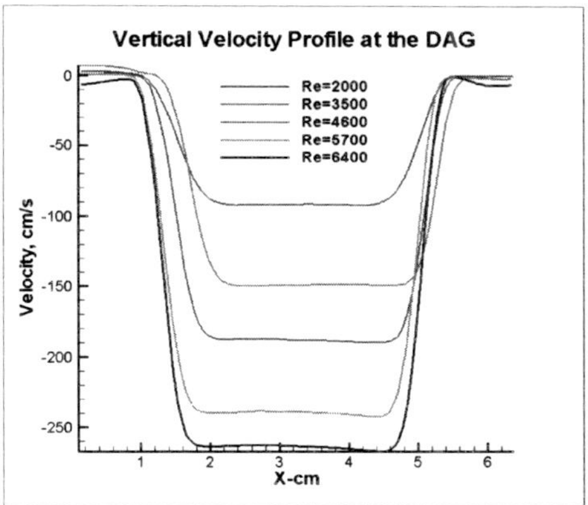

Figure 4: Velocity profile at the *DAG* of the experimental set up.

This is expected, because both concentrations at *DAG* and *RAG* are being reduced almost at the same rate. Figure 5b shows the absolute infiltration as a function of the opening height while other parameters are maintained the same. It is seen that for this case a maximum point exists at which the infiltration increases. The 2D numerical simulation for several configurations was performed by the Fluent code [20]. The simulation was performed with CO_2 as the tracer gas and validated for several test cases. Then the numerical simulation was used to increase the resolution of the required data points for the ANN, when needed. Figures 6a & 6b exhibit the streamlines and mass concentration of CO_2 as predicted by the CFD simulation. No mass flow boundary condition was specified at the *DAG*. Instead a fan boundary condition was assigned at the inner duct of the POCAC that yields the experimental flow rate. The CFD results also help us with identification of the regions in the room concerning the tracer gas sampling locations. In addition, Figure 6a indicates that the entrainment of room air occurs almost across the entire opening height and indicates that the room CO_2 is the highest near the floor.

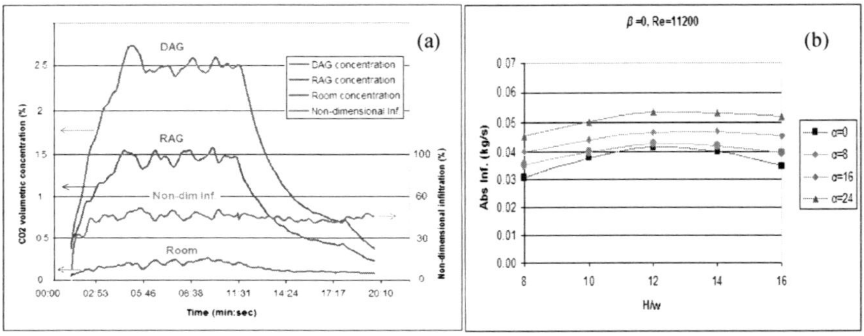

Figure 5: Experimental results: a) tracer gas concentrations and non-dimensional infiltration; b) absolute infiltration rates.

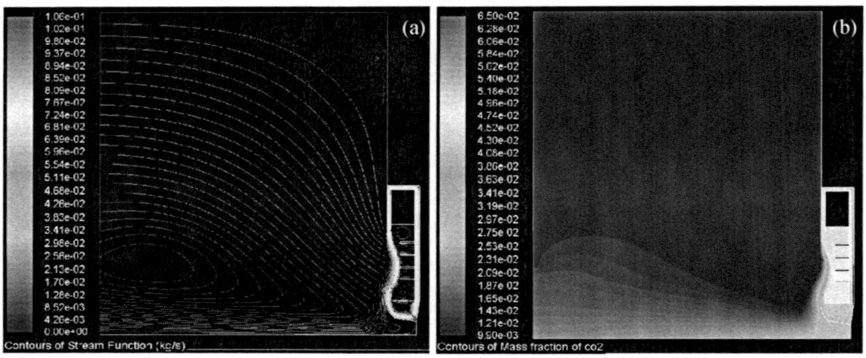

Figure 6: CFD results: a) streamline of flow inside and outside the POCAC; b) mass concentration of CO_2.

Hence positioning of the sampling probe close to the floor will produce false information regarding the room CO_2 concentration. After the completion of all tests, ANN is used as a practical tool to predict the infiltration rate for any other display case as well as conducting parametric studies. Artificial neural networks (ANNs) in fact provide an alternative paradigm or mechanism for performing regression or curve fitting as compared to classical regression methods. In other words, ANNs denote a tool that achieves an implementation of regression. The classical regression approaches are linear whereas ANN is a generalization of regression to non-linear systems. A sample of the ANN prediction after its proper training with actual data is shown in Table 2. As it is seen from this table, an excellent agreement between the prediction and actual data exists. It is evident that the accuracy of ANN prediction relies on the number of available data points (resolution).

Table 2: Comparison of the N.I.R. of experimental data with ANN predictions.

α (°)	β (°)	Re_{DAG}	$\dot{m}_{BP}/\dot{m}_{tot}$	H/w_{DAG}	Non-dimensional Infiltration Rate	
					Actual	ANN
0	-5	5500	0	8	0.239	0.244
0	0	5500	0.37	12	0.302	0.293
0	13	3400	0.46	8	0.270	0.281
16	0	3400	0.54	16	0.346	0.332
16	-5	8400	0.41	16	0.295	0.320
16	13	8400	0.41	8	0.329	0.314
24	-5	8400	0	12	0.211	0.199
24	5	2200	0.34	12	0.424	0.395
24	13	3400	1	16	0.419	0.410

6 Conclusion

A systematic approach for solving a large scale problem with many variables is suggested. The solution method relies on hybrid experimental/numerical methods and develops a practical tool based on the collected data and artificial neural network algorithm. The accuracy of results indicates the feasibility of the approach and its versatility to be applied to the problems of the same magnitude.

Acknowledgements

This work was partially sponsored by the US Department of Energy, Office of Building Technology, State and Community Programs under contract DE-AC05-00OR22725 with UT-Battelle, LLC. The authors wish to thank Southern California Edison CO. for allowing us to use their RTTC facilities for our testing and also Mr. Van D. Baxter from The Oakridge National Laboratory for monitoring the project and providing us with his support and advice.

References

[1] Howell, R.H., Van, N. & Smith, C.E., Heat and Moisture Transfer through Recirculated Plane Air Curtains. *ASHRAE Trans.,* 82(1), pp. 191–205, 1976.

[2] Howell, R.H. & Shiabata, M., Optimum Heat Transfer through Turbulent Recirculated Plane Air Curtains. *ASHRAE Trans.,* 86(1), pp. 188–200, 1980.

[3] Van N. & Howell, R.H., Influence of Initial Turbulence Intensity on the Development of Plane Air Curtain Jets. *ASHRAE Trans.,* 82(1), pp. 208–228, 1983.

[4] Rouaud, O. & Havet, M., Behavior of an Air Curtain Subjected to Transversal Pressure Variations. *J. of Env. Eng.* 132(2), pp. 263–270, 2006.

[5] Hayes, F.C. & Stoecker, W.F., Design Data for Air Curtains. *ASHRAE Trans.*, No 2121, pp. 153–167, 1969.

[6] Howell, R.H. & Adams, P.A., Effects of Indoor Space Conditions on Refrigerated Display Case Performance. *ASHRAE* 596RP, Department of Mechanical Engineering, University of South Florida, Tampa, Florida, November 1991.

[7] Stribling, D., Tassou, S.A. & Mariott D. A Two-Dimensional CFD Model of a Refrigerated Display Case. *ASHRAE Transaction*, 104, pp. 88–94, 1997.

[8] Axell, M. & Fahlen P., Vertical Display Cabinet. *SP (The Swedish National Testing and Research Institute)*: Sweden, 2002.

[9] Axell, M. & Fahlen P., Design Criteria for Energy Efficient Vertical Air Curtains in Display Cabinets. *International Congress of Refrigeration*: Washington, DC., 2003.

[10] Navaz, H.K., Faramarzi, R., Dabiri, D., Gharib, M. & Modarress, D., The Application of Advanced Methods in Analyzing the Performance of the Air Curtain in a Refrigerated Display Case. *J. of Fluid Eng.*, 124(3), pp. 756–764, 2002.

[11] Navaz, H.K., Henderson, B.S., Faramarzi, R., Pourmovahed, A. & Taugwalder, F., Jet Entrainment Rate in Air Curtain of Open Refrigerated Display Cases. *Int. J. of Refrigeration*, 28(2), pp. 265–275, 2005.

[12] Navaz, H.K., Amin, M., Srinivasan C.R. & Faramarzi, R., Jet Entrainment Minimization in Air Curtain of Open Refrigerated Display Cases, *International Journal of Numerical Methods for Heat and Fluid Flow,* 16(4), pp. 417–430, 2006.

[13] Navaz, H.K. & Faramarzi, R., Advanced Supermarket Display Case Workshop. *ASHRAE winter meeting*: Anaheim, California, January 2004.

[14] Field, B., Kalluri, R. & Loth, E., PIV Investigation of AIR-Curtain Entrainment in Open Display Cases. *IIF-II-Commission D1/B1*: Urbana, IL, 2002.

[15] Field, B. & Loth, E., An Air Curtain Long a Wall with High Inlet Turbulence. *Journal of Fluid Engineering*, 126, pp. 391–398, 2004.

[16] Chen, Y., Yuaa, X., Simulation of a cavity insulated by a vertical single band cold air curtain. *Energy Conv. & Manag.*, 46, pp. 1745–1756, 2005.

[17] Mhiri, H., Golli E., S., Berthon, A., Palec, G.L. & Bournot, P., Numerical study of the thermal and aerodynamic insulation of a cavity with a vertical downstream air jet. *Int. Comm. Heat Mass Transfer,* 25(7), pp. 919–928, 1998.

[18] Amin, M., Navaz, H.K., Dabiri, D. & R. Faramarzi, Air Curtain Studies in Vertical Open Refrigerated Display Cases: Minimization of Infiltration Rate. Final Interim Report Prepared for the Southern California Edison CO., 2008.

[19] Field B. & Loth, E., Entrainment of refrigerated air curtains down a wall. *Experimental Thermal and Fluid Science,* 30, pp. 175–184, 2006.

[20] The ©Fluent Code by Fluent, Inc.

Performance testing of reflective insulation applied in a prototype experimental chamber in Greece: experimental results for summer and winter periods

M. Gr. Vrachopoulos, D. G. Stavlas, L. D. Kravvaritis, M. K. Koukou, N. W. Vlachakis, N. G. Orfanoudakis, S. A. Mavromatis & A. G. Gonidis
Environmental Research Laboratory, Department of Mechanical Engineering, Technological Educational Institution of Halkida, Psachna, GR-34400, Greece

Abstract

With the aim of evaluating a reflective insulation's performance for utilization in the Greek climate, an experimental chamber facility has been designed and constructed at the campus of the Technological Educational Institution of Halkida, located in the agricultural area of Psachna in Evia island at Central Greece. The side-walls are a two series brick construction with a bubble material laminated between layers of aluminum foil placed in the 20mm gap of the brick layers. Results indicate that the existence of reflective insulation during the summer period averts the super heating at the interior of the experimental chamber while during winter the heat is retained in the chamber.
Keywords: reflective insulation, test room, measurement.

1 Introduction

Today, the increment of energy efficiency and the exploitation of renewable energy sources are effective ways of reducing the impacts of global warming and improving air quality while it is also an economically effective choice for consumers. The commercial and residential building sector is responsible for a large portion of the total energy consumption on a national level [1,2].

WIT Transactions on Engineering Sciences, Vol 61, © 2008 WIT Press
www.witpress.com, ISSN 1743-3533 (on-line)
doi:10.2495/HT080181

This consumption heavily concerns the covering of buildings energy demands.

In EU level the importance of energy conservation in the building sector is depicted from the Directive of the European Parliament and of the Council on the energy performance of buildings [3,4]. The Directive indicates the necessity and possibility of savings through the implementation of standard energy efficiency methods and innovative technologies, including renewable energy. In Greece, the Rule for Rational Use and Energy Savings enacted in 1998, imposes that energy consumption reduction measures should be made compulsory for all buildings by 2007 [5]. The use of energy in Greek buildings, constitutes 30% of total national energy demand and contributes about 40% of carbon dioxide emissions [6]. It should be also taken into account that climatic conditions in Greece differ from mild winters to hot summers in the south and rather cold winters to mild summers in the north of the country. These variations in climatic conditions influence in one hand the energy demand and consumption of buildings, and on the other hand the measures concerning potential and cost-effectiveness of energy conservation and those related to the exploitation of renewable energy solutions .

The peak load can be reduced by simple techniques and methods. One of them is the proper design of a building envelope and the selection of its components. The proper use of thermal insulation in buildings contributes in reducing the required air-conditioning system size and the annual energy cost. Thermal insulation is a material or a combination of materials, that, when properly applied, retards the rate of heat flow due to its high thermal resistance. For the evaluation of the performance of different insulation materials it is necessary to understand the way that heat is transferred through the composite multilayer wall construction [7,8].

In this work, aiming to the evaluation of reflective insulation's behavior for both summer and winter periods under Greek climate conditions a test room facility has been designed and constructed at the campus of the Technological Educational Institution of Halkida located in the agricultural area of Psachna at Central Greece. The side-walls are a two series brick construction with a bubble material laminated between layers of aluminum foil placed in the 20mm gap of the brick layers. Research work will be carried out through various steps. At the current first step, the outer surface of the test room wall construction is not covered with plaster. Temperature measurements have already been conducted so as to obtain a clear understanding of the reflective insulation behavior for this configuration. Experimental results show that the existence of reflective insulation during summer period deters the super heating at the interior of the test room. Due to reflective insulation's characteristics a better attenuation of temperature variations is achieved and a normal behaviour in temperature variation through the wall construction is noticed. During winter period heat remains in the test room, causing the internal temperature stabilization and the respective thermal load decrease.

2 Experimental set-up

2.1 About reflective insulation

Reflective insulation consists of one or more low emittance surfaces, such as metallic foil or metallic deposits, unmounted or mounted on substrates [9]. Because reflective insulation is usually manufactured with highly reflective, aluminum foil surfaces, 95-97% of the radiant heat that strikes the surface is reflected, and only 3-5% of the heat is emitted through the insulation. The major benefit of this reflective property is that in winter, heat inside a building is reflected off the insulation's surface back into the building so that the heat is retained inside. In the summer, heat radiated through the roof is reflected off the insulation's surface back to the roof and not inside the building thus maintaining lower internal temperatures. Unlike other insulation materials, reflective insulation reflects incident infrared radiation, thus reducing radiant heat transfer. Its performance depends on a number of factors [10] including the radiation angle of incidence on the reflective surface, temperature difference between the spaces on both sides of the reflective material, material emissivity, thickness of the air space facing the reflective material, heat flow direction.

2.2 Test room development

A small-scale test room facility has been designed and constructed on the purpose of conducting experiments under various test conditions to evaluate various thermal insulation materials and techniques, heating/cooling/ventilation technologies, etc. The test room is located at the campus of the Technical University of Halkida in the rural area of Psachna in Evia island at Central Greece (Figure 1). The dimensions of the test room are 4m x 6m x 4m and its roof is covered with roman tiles and a radiant barrier reflective insulation system.

The side walls are a two series brick construction with a bubble material lamination among layers of aluminum foil placed in the 20mm gap of the brick layers. The total wall thickness consists of 90mm brick, 10mm air gap, 1mm or 2mm reflective insulation, 10mm air gap and 90mm brick. The walls can be coated internally and externally resulting to a total width ranging from 200mm to 240mm (including a 20mm thickness of the wall sheathing-plaster on each side).

Two doors are located in the north and south-facing walls thus achieving the ventilation of the chamber. A small partition with the height of 1m is located adjacent to the north wall. Reflective insulation is a part of the test room wall construction, and more specifically: 1. Heat insulation material of the vertical wall construction to all directions (North, South, East, West), 2. Temperature and humidity insulation element of the roof.

The roof constructed of roman tiles is based on a wooden support which is protected through the application of reflective insulation and consists of the following layers: joists support of roof, wooden support with thickness 10–15 mm, cleats of reflective insulation and tiles (wooden or other material e.g. plastic) with thickness 2x10mm, and roman tile.

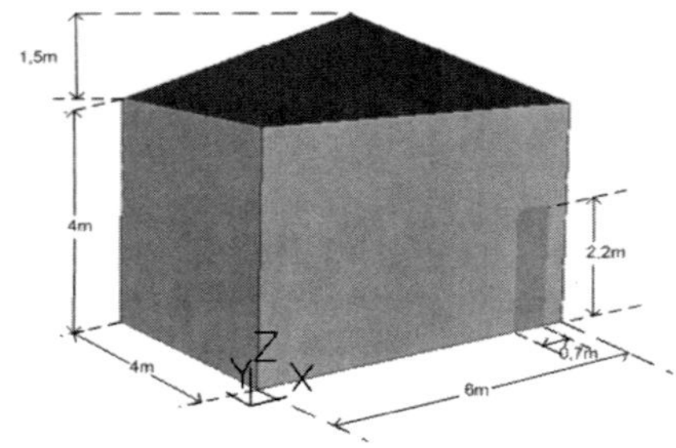

(a) (b)

Figure 1: (a) Experimental chamber; (b) geometrical details.

(a) (b)

Figure 2: (a) Reflective insulation; (b) three dimensional view of the wall construction.

2.3 Experimental plan – data acquisition

Experimental methodology is compatible with the international standards and certified methods [11,12] for the determination of the wall construction's and the other elements of the building cell, thermal behaviour. Based on the development of the periodic environmental behavior on a 24h basis and measurement analysis, the total contribution of reflective insulation to the energy operation of the wall construction (for each orientation), the solar roof and the overall behaviour of the test room can be certified.

Data acquisition has been divided into two sub-sections (Figure 3), one for temperature measurements and one for evaluation of weather conditions. Measurements have been obtained during the period February-November 2007.

The experimental equipment used for the evaluation of reflective insulation consists of temperature sensors connected to a data logging system [13,14]. The logging system has a 10 minute recording frequency. Temperatures were measured at different levels inside and outside the test room as well as inside the wall construction as shown in Figure 4. Twenty four thermocouples (six for each orientation) were connected to four modules which transmitted the signals to a RS-232 converter connected with a PC. The K type thermocouples were placed at the middle of each orientation both at the outer and internal wall.

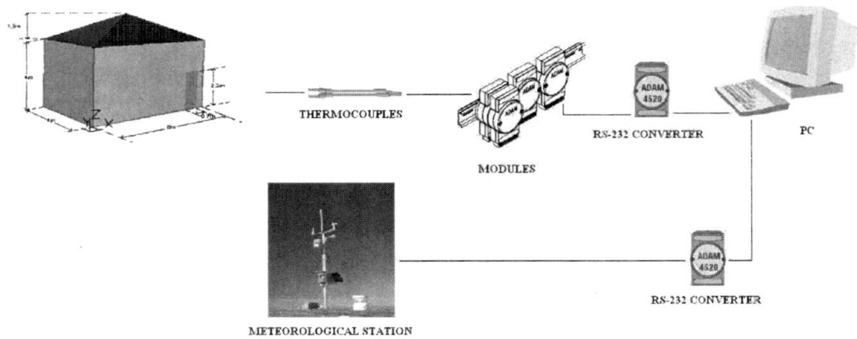

Figure 3: Data acquisition.

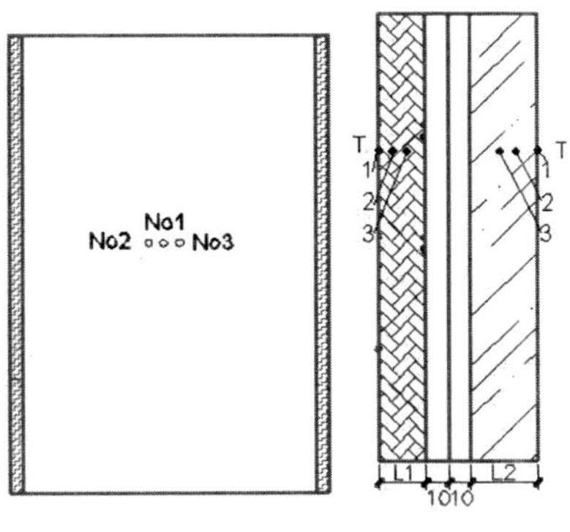

Figure 4: Temperature sensors inside the wall.

Finally, a meteorological station equipped with components and sensors that measure wind speed, wind direction, rain gauge, air pressure, temperature, relative humidity, total solar radiation and diffused solar radiation, was installed focusing on the determination of the external environment thermal behaviour. All the sensors were connected to a PC and through the use of special software measurements could be read.

3 Typical results and discussion

In this section, typical results from measurements inside and outside the test room, obtained during winter and summer periods, are presented and discussed. During the experimental period, the meteorological conditions were characterized by clear sky, quite high temperature and low wind speed.

3.1 Winter period

In Figure 5 temperature curves for the period 21-28 February 2007 and for south orientation are presented while Figure 6 shows temperature curves for a typical winter day. It can be noticed that the temperature range of the internal wall is quite smaller than of the outer one. During a typical winter day while outer wall temperature ranges from 6 to 16 °C the inner wall temperature remains quite constant from 10 to 12 °C (Figure 6). This is due to reflective insulation, which

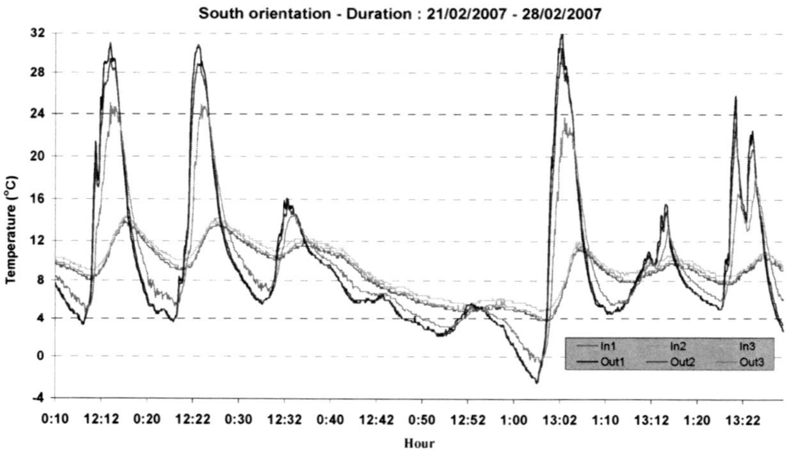

Figure 5: Temperature of inner and outer wall for south orientation – Period: 21-28/2/07.

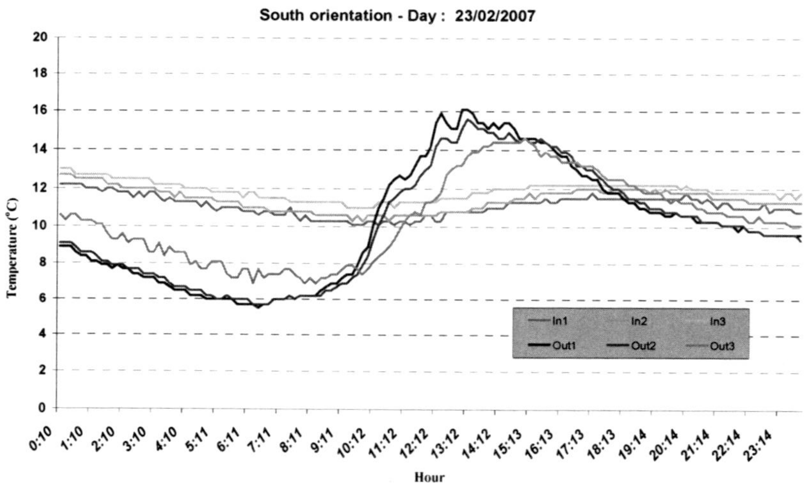

Figure 6: Temperature of inner and outer wall for south orientation – Period: 23/2/07.

during winter period permits heat reflection to the test room interior and thus heat remains in it causing the internal temperature stabilization and the respective decrease of thermal loads. The same behavior is noticed from the results of all test room orientations.

3.2 Summer period

During the summer period as is shown in Figures 7 and 8 despite external wall's high temperature variations the temperature variation at the internal walls and inside the test room is quite mild. Temperatures at the surface of the external wall of the test room ranges from 16°C early in the morning to about 32°C at 15:00-16:00 in the afternoon as it was expected. However, temperature at the surface of the internal wall of the test room is about 23-25°C during the whole day causing decrease in the required loads for the test room air conditioning. This is mainly due to the existence of reflective insulation which during the summer period deters the super heating at the interior of the test room. When solar radiation falls on to a non-transparent surface, like the test room walls, the outer surface absorbs a part of the radiation which is transformed to heat. A part of the heat is re-emitted to the outside while the rest is driven through the wall to the test room interior with a rate depending on the thermal characteristics of the wall construction materials. Because of the reflective insulation characteristics, a better attenuation of temperature variations is achieved and a normal behavior in temperature variation through the wall construction is noticed. This is due to the time duration between the peak of the outside temperature (about 19:00) and the peak of the inside surface temperature (about 21:00). Actually, a time delay of two hours means that the test room wall construction materials will start to reject heat to the test room interior during the night (when there is a demand for heating during the winter or when room cooling is feasible through ventilation

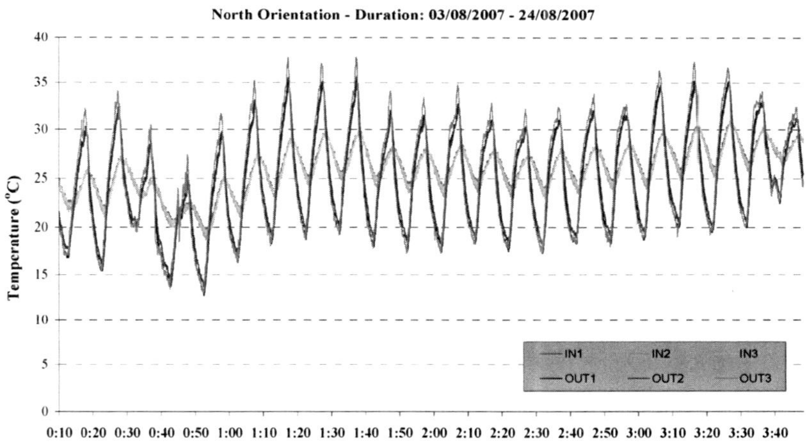

Figure 7: Temperature of inner and outer wall for north orientation – Period: 3-24/8/07.

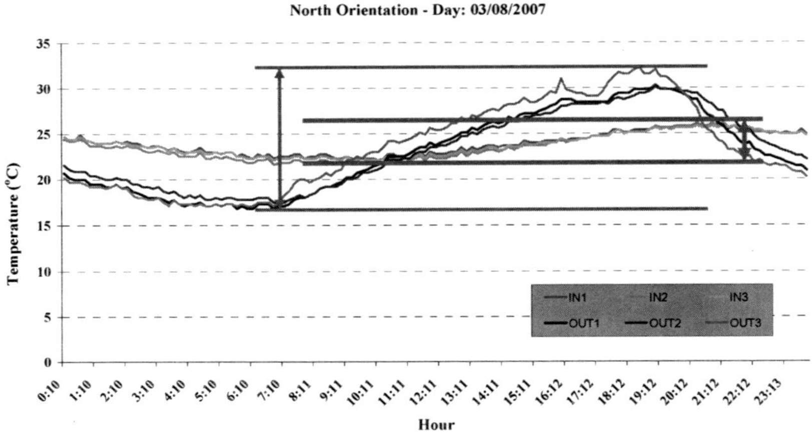

Figure 8: Temperature of inner and outer wall for north orientation – Period: 3/8/07.

during summer). In light constructions, there is a small time delay and thus outside temperature variations become tangible at the building interior quite quickly. During the night, air temperature at the test room interior is higher in comparison with the outside temperature and there is a heat flow to the outside causing the decrease of the wall temperature together with the temperature at the room interiors. Consequently, at the next day, the room becomes quite cool so as to be able to absorb again the undesired heat. The same behavior is noticed from the results of all test room orientations.

4 Conclusions

In the present work, the possibility to exploit reflective insulation in Greek climate has been investigated focusing on the improvement of building insulation behavior. For that purpose, an experimental chamber facility has been designed and constructed at the campus of the Technological Educational Institution of Halkida, located in the agricultural area of Psachna in Evia island at Central Greece. Results show that during winter period reflective insulation permits heat reflection to the test room interior and thus heat remains in it causing the internal temperature stabilization and the respective decrease of thermal load. During the summer period, reflective insulation deters the super heating at the interior of the test room, a better attenuation of temperature variations is achieved and a normal behavior in temperature variation through the wall construction is noticed.

Acknowledgement

This research work was co-funded by the European Social Fund (75%) and National resources (25%) through the Operational Program for Educational and Vocational Training II (EPEAEK II) "Archimedes II".

References

[1] IPCC 1996. Technologies, Policies and Measures for Mitigating Climate Change – IPCC Technical Paper 1; Geneva: Intergovernmental Panel on Climate Change; 84pp.
[2] Zografakis, N., Technologies for rational use and savings of energy in buildings. Energy (in Greek), 62, pp. 112–114, 2000.
[3] Directive 2002/91/EC of the European Parliament and of the Council of 16 December 2002 on the energy performance of buildings, 4/1/2003, Official Journal L1 65.
[4] Chwieduk, D., Towards sustainable-energy buildings. Applied Energy, 76, pp. 211–217, 2003.
[5] Joint-Ministerial Decision 21475/4707 for the reduction of CO2 emissions by means of energy efficiency improvement in buildings. Government Newspaper, No. 880, 19th August 1998.
[6] Regulating Authority for Energy (RAE), The Energy System in Greece [in Greek] (http://www.rae.gr/energysys/main.htm).
[7] Incropera, F.P. & DeWitt, D.P., Fundamentals of Heat Transfer, 4th Ed., John Wiley & Sons, New York, 1996.
[8] Bird, R.B., Stewart, W.E. & Lightfoot, E.N., Transport Phenomena, John Wiley & Sons, New York, 1960.
[9] ASTM C1224-03 Standard Specification for Reflective Insulation for Building Applications, ASTM International, 2005.
[10] American Society of Heating, Refrigerating, and Air Conditioning Engineers (ASHRAE). Handbook of Fundamentals, Atlanta, GA, USA, 2001 [Chapter 23].
[11] ASTM C177-04 Standard Test Method for Steady-State Heat Flux Measurements and Thermal Transmission Properties by Means of the Guarded-Hot-Plate Apparatus, ASTM International.
[12] ISO 7726, Ergonomics of the thermal environment – Instruments for measuring physical quantities, International Organization for Standardization, Geneva, Switzerland, 1998.
[13] Supco DLTH Temperature/Humidity Data Loggers, Instruction manual, USA, 2002.
[14] Supco LOGiTpc Interface Software, User's Guide, USA, 2002.

A measuring method based on photodiodes for the diagnostic of optimal combustion conditions

L. Arias[1], O. Farías[2], S. Torres[1] & D. Sbárbaro[1]
[1]*Department of Electrical Engineering, University of Concepción, Chile*
[2]*Department of Mechanical Engineering, University of Concepción, Chile*

Abstract

A non-intrusive method for monitoring flames in gas burners is presented. The method is based on the optical analysis of the flame by using a Silicon Photodiodes array and a set of interference optical filters. The covered wavelengths evaluate the formation and behaviour of excited CH* and C_2* radicals for confined gas flames. The monitoring of these radicals is carried out in the flame reaction region at the burner exit, where the C_2*/CH* ratio signals provide enough information to identify the optimal fuel–air ratio in order to obtain the boiler maximal efficiency and to prevent an incomplete combustion. The experimental results obtained for two power levels, in a boiler of 150 kW, demonstrate that the C_2*/CH* ratio provides important information concerning the combustion state and gives a clear indication about the burner adjustment when CO emissions begin to increase to the upper the levels accepted by environmental laws. The fast response and its non-intrusive character give to the proposed optical sensor an important potential to be used in advanced control strategies for the on-line optimization of the combustion process.
Keywords: combustion, gas, flame spectrum, photodiodes, optimization, optical sensor.

1 Introduction

The monitoring of the combustion state in boilers and industrial furnaces is limited due to the hostile environment. Indeed, the high temperature, the corrosive atmosphere and the limited access to the flame characteristics restrict a reliable diagnostic. The information provided by conventional combustion

WIT Transactions on Engineering Sciences, Vol 61, © 2008 WIT Press
www.witpress.com, ISSN 1743-3533 (on-line)
doi:10.2495/HT080191

instruments is clearly insufficient to act on the burner settings, which would be an effective way to optimize the flame, to operate at high thermal efficiency, reducing emissions of carbon monoxide and/or nitrogen oxides. Thus, the development of wireless sensors with a non-intrusive character is highly important.

The application of control techniques for combustions systems is of great interest. However, reliable combustion control techniques need reliable sensors and this depends on the characteristics of the combustion process. In this case, many groups of sensors has been analysed and applied at an industrial level [1,5]. For gas sensing, the following classes of sensors have been revised:

- Solid electrolyte ZrO_2 sensors for O_2 and CO.
- Solid electrolyte, non-nernstian ZrO_2 sensor for CO/H_2.
- Thin-film semi-conductor sensors (SnO_2, Ga_2O_3, Nb_2O_5) for oxidising or reducing gases.
- Thermal conductivity sensor for CO_2.
- Sensors and optical methods using radiative properties of flames: spectrometry, spatio-temporal signature.
- The potential of some generic technologies being developed, in particular those using laser diodes and those using diamond coatings.
- Optical methods using the flame signature principle.

The optical methods, used in the characterization of combustion quality and pollutant emissions by means of the flame signature principle, have been studied and developed in many works. However, their application at an industrial level is difficult today [1,5].

In particular, the massive use of radiometric techniques [3,4,7,10,11,22] has contributed to a better understanding of the combustion processes. It is known that in the reaction zone of hydrocarbon flames, the oxygen of the air reacts with the fuel to liberate the chemical energy stored in the molecular bond. This energy is reflected in the visible flame spectrum around certain narrow wavelengths and a continuous background. The discontinuous spectra of atoms or molecules are arranged in groups, each group being called a band of free radicals [2]. A typical case of gas flames burned by diffusion into the surrounding air show yellow luminosity due to the formation of soot. Soot emissions are similar to that of a grey body but with a more complex variation of emissivity with the wavelength [3,12]. In gas premixed flames, a CO continuum has been identified in the range of wavelength from 300–550 nm [3].

The radical's intensities detection has been greatly used to obtain the combustion state index. The radicals better detected are those found in the short wavelength, because of having a lower interference with the emission of soot particles. In this case, CH* (at 432nm) and C_2* (at 516nm) radicals are very interesting because they provide easily observable signals, with a strong intensity, and their radiation is spectrally resolved. OH* radical (306-315nm) is attractive because of the reduced background emission at this wavelength. On the other hand, the CH* radical has been identified as the best marked on the flame front and it is often used as a measure of the heat release rate. In many works, the intensity emissions and the ratio of different radicals have been used in order to identify clearly the main flame zones, air–fuel ratio, and pollutant emissions amongst others. Thus, the ratios CH*/OH*, C_2*/OH* and C_2*/CH* show a nearly linear function of equivalence ratio [19,22]. Also, the correlation between chemiluniscence of any radicals and the flame temperature has been investigated [22].

The study and application of sensitive photodiodes in the UV-VI range, for spectral emission detection, was carried out in the last 10 years. However, the polarisation mode of these photodetectors is not mentioned. Khesin [14,15] applied photodetectors to realize combustion control in pulverized coal burners at an industrial level, by measurement of the flame instability. In recent years a method based on a Si photodiode (with spectral response 190-1100nm) has been used [17,18]. The signal of this sensor integrates the contribution of different excited radicals and continuous spectra and uses the rms average of the 'flame luminosity', which is compared with the CO emissions. Also, a Photomultiplier has been used with an optical filter in 310 ± 10nm to detect the OH* radical intensity. Another method is based on the utilization of UV cells [16], which is based on Si (UV enhanced) and GaP junctions and the signal generated by the photodetectors is acquired and post-processed. The light is collected by means of a lens and optical fibre and the detector cell is installed in a tube, inside the burner. A correlation has been obtained between O_2 emissions and the flame stability, but this correlation changes for different boiler geometry, burner/boiler types and load.

Additionally, others techniques have been used by means of the flame spectrum. For example, the flame temperature has been estimated by using the two or more wavelengths methods, through the spectral emission measurement of soot particles [2,3,9,11]. Moreover, the gas emission has been estimated through absorption spectroscopy by using laser diodes [1,5,6]. On the other hand, different camera technologies have been used to define combustion parameters from the flame morphology [19,21].

The flame spectrum intensity depends on many variables such as the temperature, the kind of gas mixture, fuel–air ratio, the burner type, the turbulence, and the flame area observed. Therefore, the main purpose of this investigation is to characterize the influence of the combustion parameters by use of a photodiode array in order to propose an optimal combustion control system. Thus, an optical method for monitoring flames in gas burners is developed. Further, the optical analysis of the flame is achieved by using a Silicon Photodiodes array and a set of interference optical filters. The optical sensor is able to measure the C_2* and CH* signals, and the ratio of these radicals provides enough information to identify the optimal fuel–air ratio in order to obtain the boiler maximal efficiency and to prevent an incomplete combustion.

2 Monitoring a gas flame by using photodiodes

The objective of this paper is to describe the advantage of the photodiodes utilization in order to determinate the combustion state and use this information for a combustion control strategy.

In this paper, a new monitoring method for CH* and C_2* radicals, by using silicon photodiodes equipped with optical filters centered in the corresponding emissions bands is presented.

The spectral responsivity of these devices is between 380–1200nm, with a maximal response at 850nm. The center wavelength of filters for detection of CH* emissions is 430 ± 5 nm and 514.5 ± 5 nm for C_2* emissions, with a transmission of 45% and 70%, respectively. The difference between the filter transmission and responsivity intensity was corrected by software. The polarization mode for driving the photodiodes signals was the Photoconductive. Thus, a linear response is obtained by using this mode between the photon incidence and the current generated.

The flame area covered by the photodiodes is observed from the exit burner. The specific installation of these devices is shown in figure 1, where the photodiode array is cooled by means of fresh gas and air flux and the distance between the photodiodes and the flame basis is 8cm. The reaction zone is observed, with a FOV of 72°, where the formation of these radicals is strong [3,22]. The experimental results were performed in a central heating boiler of 150 kW, equipped with a Riello GS20 gas burner. The test bench was instrumented with gas and water flow meters, thermocouples and gas analyser (at the exit of the chimney) in order to obtain the on-line parameters of combustion and boiler efficiency. The photodiodes driver's signals were recorded in the computer by using a DAQ board at 100 kHz, and processed by the LabView interface. The tests reported in this paper were performed covering two firing rate values: 150 and 130 kW.

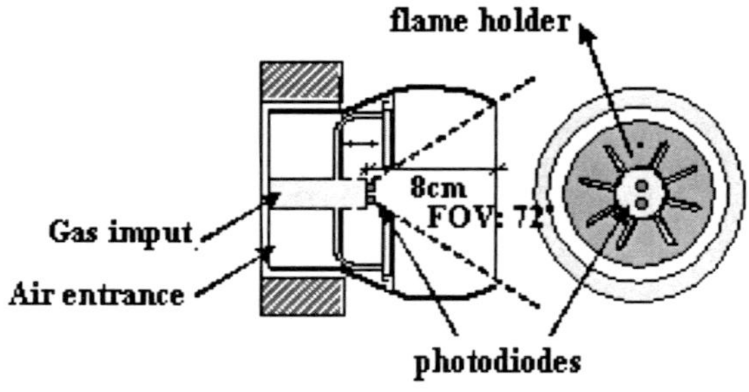

Figure 1: Photodiodes installation.

3 Experimental results

Figure 2 show the spectral intensity of the gas flame, obtained from the burner exit by using the USB2000 Radiometers operating in the visible range. It is possible to distinguish clearly the main bands of CH* and C_2* radicals. One verifies that the main reacting zone may be characterised by these specific emissions and provides more observable signals to be monitored by the photodiode array.

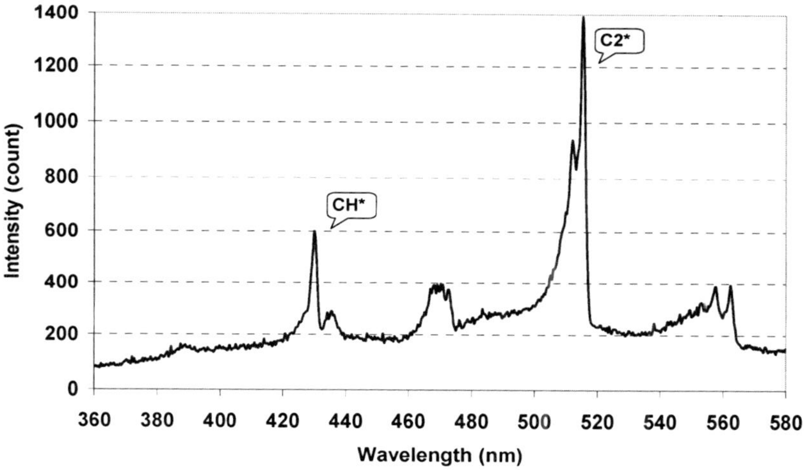

Figure 2: C_2* and CH* radicals at the reaction zone.

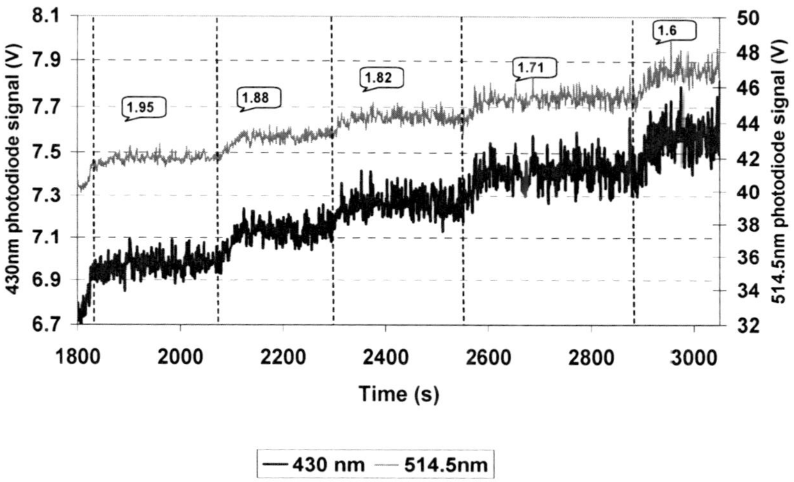

Figure 3: On-line response, photodiodes at 430 and 514.5nm.

In the tests performed, the air flow was decreased by closing the burner air damper and the real time response of the photodiodes and gas analyzers were registered. Figure 3 shows the online response of the photodiodes, for different air excesses (Lambda=1+e), where it's possible to observe that the change in the air flow is followed by the array sensor. Also, as in figure 2, the C_2* radical intensity is stronger than the CH* radical. The great difference between the

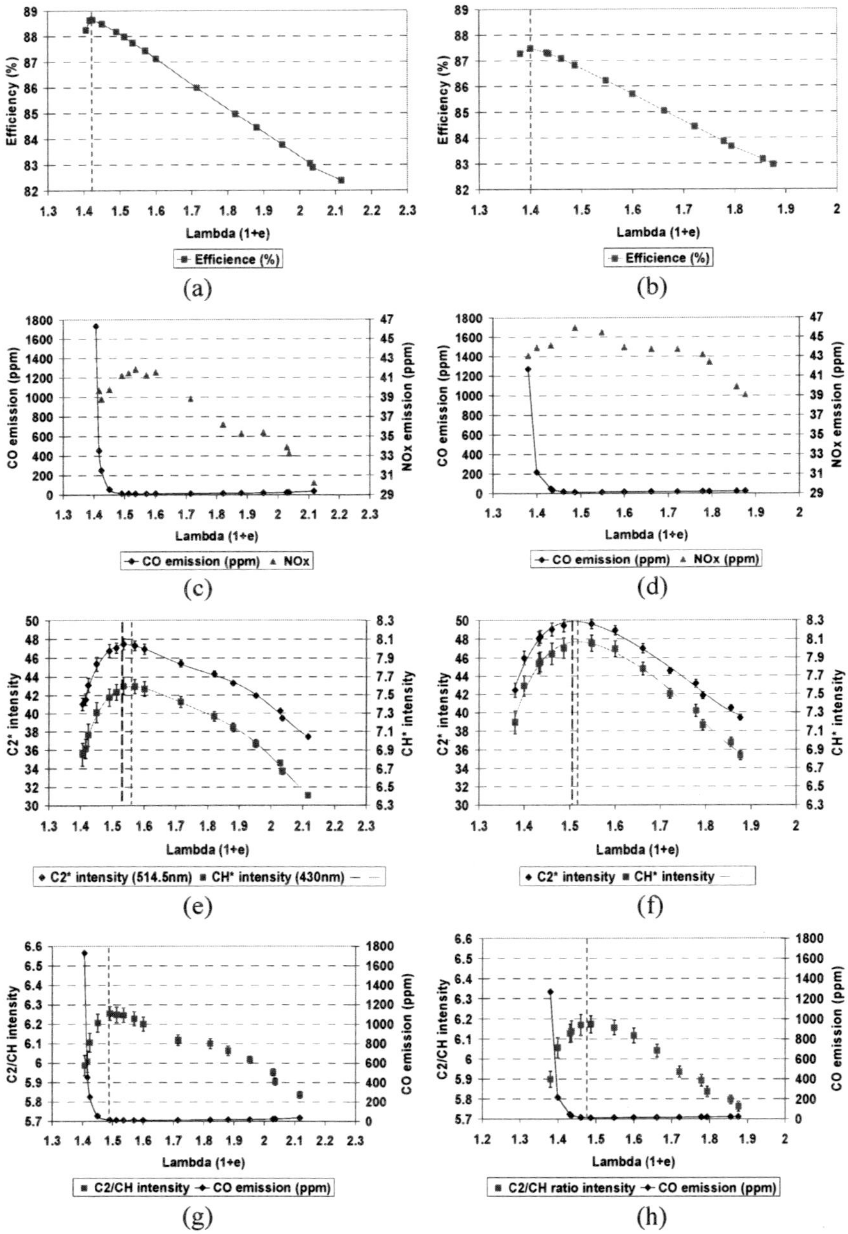

Figure 4: Test boiler resume for two power levels: 130 kW (in left column) and 150 kW (in right column) for change in the air excess. (a,b) Thermal efficiency (c,d) CO and NOx emissions. (e,f) Absolute intensity of CH* and C_2* radicals detected by the photodiodes array. (g,h) C_2*/CH* intensity ratio.

intensity of these signals is because the filters have a FWHM (full width at half maximum) of 10 nm. In this case, the integrated signal from the photodiode at 514.5 nm is bigger than photodiode at 430 nm

Figure 4 shows the results of the boiler performance and the radical emissions for two levels of boiler power (130 and 150 kW), as a function of the excess air. The optimal operation condition is clearly given by critical excess air when CO emissions increase abruptly and the boiler efficiency began to decrease at low levels of air excess, as show in figure 4 (a, b, c and d).

From the point of view of the pollutant emissions, the trends of NOx curves (figure 4, c and d) are very similar to the radical emissions (e and f). The intensity emissions of both radicals, CH* and C_2*, show a similar behaviour. The maximal levels are achieved at excess air values higher than the critical one defined by increased CO. However, these maximal levels shows a slight diphase for CH* and C_2* radicals, according to the observations of Gaydon and others [8,3].

In figure 4 (g and h) one can observe the trends of the C_2*/CH* intensity ratio, which increases while excess air decreases to a level near to the critical point of incomplete combustion, i.e. when CO increases abruptly.

4 Conclusions

The experimental result shows that the C_2*/CH* intensity ratio is a good indicator of the combustion state, because it permits prediction of the point of optimal air excess condition in order to obtain the maximal efficiency of the equipment, assuring a low level of pollutants (CO and NOx). This fact has been verified for two power conditions of a gas boiler. From this result it is possible to perform an optimal strategy of combustion control.

The fast response, non-intrusive character, and low cost yield to the developed optical sensor is an important potential to be used in advanced control strategies for the on-line optimization of the combustion process.

References

[1] Michel J. B., Champinot C., and Dugue J. *State of the art on emerging combustion control sensors*, 6th Int'l Conf. on Tech. and Combustion for a Clean Environment, Oporto, 2001.

[2] Zizak, Giorgio. *Flame Emission Spectroscopy: Fundamentals and Applications*, CNR-TeMP, Instituto per la Tecnologia dei Materiali e dei Processi Energitici Via Cozzi 53, 20125 Milano, Italy, 2000.

[3] Farias, O. *Towards the Development of an Optimal Combustion Control in Fuel-Oil Boilers from the Flame Emission Spectrum*, Thèse de Doctorat, Faculte des Sciences Appliquees, Universite de Liege, Belgique, 1997.

[4] Ikeda, Y., Kojima J., and Nakajima, T. *Fast Response Local Equivalence Ratio Measurement in Premixed Turbulent Flame*. Department of Mechanical Engineering, Kobe University, Japan.

[5] Docquier N., Candel, S.: "Combustion control and sensors: a review", Laboratoire EM2C, Ecole Centrale Paris, CNRS UPR 288, Grande Voie des Vignes, F-92295 Châtenay-alabry Cedex, France, 2001

[6] Webber, M., Wang, J., Sanders, S., Baer, D. and Handson, R. *In situ combustion measurements of CO, CO_2, H_2O and temperature using Diode Laser Absorption Sensors*, High Temperature Gasdynamics Laboratory, Department of Mechanical Engineering, Standford University, Standford, CA 94305-3032, USA.

[7] Romero, C., Li, X., Keyvan, S. and Rossow, R. *Spectrometer-based combustion monitoring for flame stoichiometry and temperature control*, Energy Research Center, Lehigh University, Bethlehem and Center for artificial Intelligence in Engineering and Education, University of Missouri, Columbia, USA, 2004.

[8] Gaydon A.G. and Wolfhard H.G., *Flames: their structure, radiation and temperature,* Chapman and Hall Ltd, 1970.

[9] Bar-ziv, E. *Thermal Radiation and Optical Pyrometry*, Department of Mechanical Engineering Ben-Gurion University, Beer-Sheva, Israel, 2003.

[10] Farias, O., Ngenndakumana P. *Diagnóstico de la combustión, a partir del espectro de llamas luminosas*, Revista de Ingeniería, Universidad de Concepción, Chile. Año 10, Edición N°1, Noviembre 1998.

[11] Ngendakumana P., Zuo B. and Winandy E., *A spectroscopic study of flames for a pollutant formation regulation in a real oil boiler*, 2nd International Conference on Combustion Technologies for a Clean Environment, Lisbon, 1993.

[12] Pearse R. W, Gaydon A.G *The identification of Molecular Spectra*, Chapman and Hall, London, 1965

[13] Adams W.H. *Heat transmission*, Mc Graw-Hill Publishing Company Ltd, London, 1954.

[14] Khesin, M. *Demonstration tests of new burner diagnostic system on a 650 MW coal-fired utility boiler*, American Power Conference, 59th Annual meeting, 1997.

[15] Khesin, M. *Demonstration of new frequency-based flame monitoring system*, Proc. 59th Ann. Mtg. of the American Power Conf., Vol: 58–II, pp. 1010–1013, 1997

[16] Michel, J., Chetelat, O. *Flame Signature as a Low-Cost Flame Control Method*, 5th Int'l Conf. on Tech. and Combustion for a Clean Environment, pp. 1251–1253, 1999

[17] Ballester, R., Sanz, A. *Advanced monitoring and control of burners using flame sensors*, Fluid Mechanics Group / LITEC, University of Zaragoza, Spain, 2006.

[18] Ballester, R., Sanz, A. *Identification of combustion conditions using flame sensors*, 2005 AFRC Intl. Symposium on Dynamics and Control of Industrial Combustion Processes, Georgia Institute of Technology

[19] Martins C., Carvalho J., Ferreira M.: *CH and C_2 radicals characterization in natural gas turbulent diffusion flames*, Journal of the Brazilian Society of

Mechanical Sciences and Engineering Print ISSN 1678-5878 J. Braz. Soc. Mech. Sci. & Eng. vol.27 no.2 Rio de Janeiro Apr./June 2005

[20] Kauranen P., Aandersson-engels S. *Spatial mapping of flame radical emission using a spectroscopy multi-color imaging system*, Department of Physics, Lund Institute of Technology, Lund, Sweden, 1991

[21] Sbárbaro D., Farias O. *Real time monitoring and characterization of flames by principal-components analysis*, Department of Electrical and Mechanical Engineering University of Concepcion, 2002.

[22] Morrel M., Seitzman J. *Interpretation of optical emissions for sensors in liquid fueled combustors*, AIAA-2001-0787 at the 39th AIAA Aerospace Sciences Meeting, Reno, NV, January 8–11, 2001

Author Index